Synthese Library

Studies in Epistemology, Logic, Methodology, and Philosophy of Science

Volume 531

The aim of *Synthese Library* is to provide a forum for the best current work in the methodology and philosophy of science and in epistemology, all broadly understood. A wide variety of different approaches have traditionally been represented in the Library, and every effort is made to maintain this variety, not for its own sake, but because we believe that there are many fruitful and illuminating approaches to the philosophy of science and related disciplines.

Special attention is paid to methodological studies which illustrate the interplay of empirical and philosophical viewpoints and to contributions to the formal (logical, set-theoretical, mathematical, information-theoretical, decision-theoretical, etc.) methodology of empirical sciences. Likewise, the applications of logical methods to epistemology as well as philosophically and methodologically relevant studies in logic are strongly encouraged. The emphasis on logic will be tempered by interest in the psychological, historical, and sociological aspects of science. In addition to monographs *Synthese Library* publishes thematically unified anthologies and edited volumes with a well-defined topical focus inside the aim and scope of the book series. The contributions in the volumes are expected to be focused and structurally organized in accordance with the central theme(s), and should be tied together by an extensive editorial introduction or set of introductions if the volume is divided into parts. An extensive bibliography and index are mandatory.

Pedro Atã • João Queiroz

Mind as an External Sign: A Semiotic Externalist Philosophy of Cognitive Science

Springer

Pedro Atã
Free State University
Bloemfontein, South Africa

João Queiroz
Universidade Federal de Juiz de Fora
Juiz de Fora, Brazil

ISSN 0166-6991 ISSN 2542-8292 (electronic)
Synthese Library
ISBN 978-3-032-08987-8 ISBN 978-3-032-08988-5 (eBook)
https://doi.org/10.1007/978-3-032-08988-5

This Springer imprint is published by the registered company Springer Nature Switzerland AG
The registered company address is: Gewerbestrasse 11, 6330 Cham, Switzerland

Contents

Chapter 1
Semiotic Turn in Distributed Cognition

Mind is sign action—an emergent process of cognitive niche construction. This thesis on cognitive externalism is based on Peirce's pragmaticism and semiotic theory of mind. The recent history of cognitive externalism as a research domain includes approaches such as the extended mind (Clark and Chalmers 1998), situated problem-solving (Kirsh 2009), distributed cognition (Hutchins 1996), the dynamical systems approach (Port and Van Gelder 1995; Chemero 2011), socio-cultural niche construction (Sinha 2024), and embodied and enactive cognition (Gallagher 2023). Some of the central questions investigated in this domain have to do with the nature of representations—which give rise to positions such as distributed representationalism, as well as anti-representationalism—and the relation between cognitive artifacts and agents—which gives rise to positions such as agent-centric externalism and distributed externalism.

We argue for a semiotic turn in distributed cognition. The central problems of externalist cognitive science should be examined from the perspective of sign action. What is meaning in external cognition? How does it emerge? How to model the external action of signs? How to conceptualize the relation between artifacts and agents in problem solving? In building our semiotic approach to cognitive externalism, we are especially influenced by recent discussions on emergentism, multi-level systems, situated problem-solving, niche construction, and process ontology. Our position within this scenario is to advocate for a semiotic turn in cognitive externalism, based on C.S. Peirce's thesis that mind is sign action (semiosis). Attempts very close to what we will do here have produced results within the scope of what has been called the pragmatist turn in cognitive science (Bernstein 2010; Engel et al. 2016).

Peirce's pragmatism can be understood as a philosophy of temporal distribution, in which any being exists in relation to the future effects it produces, as derived from the past interactions it has had. In this philosophical framework, forms, ideas, and concepts exist only and always through their dynamic instantiation in effects. As

P. Atã, J. Queiroz, *Mind as an External Sign: A Semiotic Externalist Philosophy of Cognitive Science*, Synthese Library 531,
https://doi.org/10.1007/978-3-032-08988-5_1

pragmatism is a philosophy of process and action, Peirce's semiotics is action-oriented. In Peirce's philosophy, the mind is a semiotic process, and semiosis is a kind of action, a mode of producing effects. Thinking is sign action, beliefs are dispositions for action, and inquiry is a process whose objective is to fix dispositions for action. A reader of new developments in Philosophy of Cognitive Science will recognize these ideas. Peirce's cognitive semiotics anticipates the pragmatic turn in 4E cognition. However, and surprisingly, little attention has been dedicated to Peirce's pragmatism in cognitive externalism, compared to Dewey (Gallagher 2009), James (Jung and Madzia 2016), and philosophers in the phenomenological tradition, such as Heidegger (Kiverstein and Wheeler 2012; Wheeler 2007) and Merleau-Ponty (Dreyfus 2002; Chemero 2011).

A semiotic turn in distributed cognition asks us to understand artifacts, problem-solving tasks, and cognitive niches as semiosis (sign-in-action). The word 'sign', here, refers not only to the physical artifacts as such, but to the artifacts as embodiments of sign processes. Signs tend to grow towards a state of relative stability, to reproduce, and to reproduce the conditions of their own reproduction. Cognitive artifacts are signs and they are not physical artifacts manipulated by an agent. Rather, agents and artifacts are instantiated parts of an emergent self-organized process that reproduces itself further and adapt towards conditions of stability (niche construction). Or, to be more precise, the physical artifacts are instantiations of a process of interpretant production and growth of habits, which recruits not only the physical artifacts themselves, but entire distributed cognitive systems. A semiotization of distributed cognition involves overcoming any cartesian psychology dualism (see Wheeler 2007) between agents and artifacts.

When we semiotize cognitive artifacts, we cease to see them as mere tools submitted to the purposes of an agent, or group of agents, and instead we see them as animated by an internal dynamic of growing habits throughout various instantiations. Each instantiation reproduces the sign again in new conditions. Thus, to say that signs are external, in our approach, involves three implications: (i) signs are processes, distributed in time across several instantiations, which cannot be located wholly in any present (temporal distribution); (ii) signs cannot be seen merely as tools for the execution and expression of goals and decisions made internally by an agent; (iii) signs are self-organized and emergent.

This is a materialist perspective in the sense that it doesn't treat signs as extensions of ideas and motivations that are inner and psychological. However, it is also a non-physicalist perspective (in the sense of reductive physicalism) since it sees these processes as inextricably connected to habits, which here are not ideal forms, but conditions of stability, temporally distributed, and thus enmesh actualities with possibilities and general tendencies. Processes such as authorship (see Chap. 10), poetry improvisation (Chap. 7), and problem solving (Chaps. 5 and 6) are understood as material processes, with spatial and temporal situatedness, with enabling and boundary conditions, as well as costs.

1.1 Sections and Chapters

1.1.1 First Part: Semiotic Emergence

We describe semiosis as an emergent process. This is the more methodological part, where we present philosophical assumptions, theoretical tools, and the most general properties of semiosis as an emergent process. We also illustrate the model with a historically well-known case—the emergence of Classical ballet.

1.1.1.1 Chapter 2: Active Semiotic Externalism

The chapter explores Peirce's cognitive semiotics, which conceives cognition as sign action, or semiosis—an active, embodied, situated, and distributed process. This view, termed active semiotic externalism, encompasses representation, dynamic situated relations, and artifact manipulation. Peirce's framework models semiosis as a triadic relation within a theory of relations, asserting that relations are explanatorily more fundamental than entities. This perspective aligns with Peirce's metaphysical foundationalism and realism about relations. Peirce's categories, influenced by Aristotle, Kant, and Hegel, offer a systematic understanding of relations and serve as rudiments of experience. The triadic hierarchical process of semiosis models the self-organization of signs—how signs generate further signs in a distributed manner across time and space. Peirce's triadic system of categories serves as a thinking tool, encapsulating the central theses of cognitive semiotics and offering a conceptual model for understanding semiosis and cognition. It allows for the identification, classification, and analysis of semiotic phenomena.

1.1.1.2 Chapter 3: Semiosis as an Emergent Process

Theories of emergence describe qualitative differences between levels or domains of a process, where emergent properties cannot be deduced from constituent parts. Emergentism suggests that reality operates across distinct levels with unique rules, explaining true novelty within a naturalistic view. In semiotics, emergence explains how sign processes possess unique properties that depend on base-level interactions but cannot be reduced to them. It requires a naturalistic stance while acknowledging semiosis' creative capacity to constitute epistemological and/or ontological domains with unique rules. Temporal distribution is crucial, operating across different temporal scales. Emergence aligns with a process-oriented ontology, focusing on how organized processes hold causal powers, explaining how sign processes grow across multiple instantiations. It can be understood diachronically or synchronically, emphasizing temporal transformation or simultaneous continuity and discontinuity. Weak and strong emergence theories distinguish between explanatory/epistemological and ontological phenomena. Semiotic emergence can be hierarchical or flat/

transformational. In a hierarchical sense, semiosis involves nested relations among signs, objects, and interpretants, with a hierarchy of presuppositions and constraints. In a flat sense, semiosis is a dynamic process producing interpretants successively, influenced by complex conditions, highlighting the dynamic nature of meaning production.

1.1.1.3 Chapter 4: The Emergence of Sign Action: The Case of Classical Ballet

In this chapter, we model a case of semiotic emergence—the history of ballet. Ballet, as an artistic tradition, emerges through the interplay of different initiating and boundary conditions, resulting in the establishment of regular tendencies (habits) operating between dancers, spectators and performance spaces. Throughout the sixteenth to the nineteenth centuries, ballet underwent various changes, including the codification of dance steps, alterations in performance spaces, and standardization of training techniques, leading to reduced variability and increased redundancy in its sign action. When ballet transitioned from court halls to the proscenium arch stage, it favored a frontal position of observation, creating a niche for more vertical movement styles, a trend aligned with the emphasis on ethereal characters like sylphides in literary romanticism and incentivized the development of new artifacts such as pointe shoes. In this description, semiosis is an emergent process of development of artifacts.

1.1.2 Second Part: Problem Solving Tasks

We describe semiosis as a problem-solving process. Here, the focus is on semiotizing cognitive externalism. We do this by modeling externalist cognition as a triadic interaction between artifacts, problems, and solutions.

1.1.2.1 Chapter 5: Artifacts in Problem Solving: Iconicity and Abductive Inference

Within the context of problem-solving, artifacts serve as embodiments of problems. In external cognition, a problem is presented to an agent. This presentation requires artifacts. It is only through artifacts that an agent can engage with a problem (act towards a goal). However, there are different ways in which artifacts may embody problems for an agent. Isomorphic representations of the same problem illustrate this. In particular, the rules that bound a problem may be presented to an agent either in the form of declarative propositions or in the form of material, artifactual constraints. When artifacts embody the rules for an agent (externalization of constraints), they are icons, whose object is a problem space bound by rules.

1.1.2.2 Chapter 6: The Object of Iconic Artifacts: Why an Underground System Is a Semiotic Problem

Beyond transforming how the rules of a problem are presented to an agent, an artifact can transform the problem itself. For example, the network diagram of a subway system transforms the task of navigating the subway, in comparison to a geographically accurate map. When an artifact transforms the problem, it acts as a cognitive niche that allows that problem to be acted upon in some way. Thus, the diagram of the subway constructs the subway as a semiotic object (not merely an existent). Just as there cannot be a cognitive task (problem) without an artifact that embodies it, there cannot be a semiotic object without a sign that grants it its objecthood.

1.1.2.3 Chapter 7: Poetry Improvisation as Problem Solving in Distributed Cognitive Systems

Artifacts and problems co-determine each other in a continuous process. If we describe externalist cognition merely as the use of artifacts to solve problems efficiently, we lose sight of the process itself that mediates between artifacts and problems, defining them in relation to one another. For example, *repente*, a tradition of improvised oral poetry in Brazil, is characterized by the problems that poet-singers must solve (improvisation under several constraints) and by the artifacts they learn to master in order to solve those problems (versification patterns, musical instruments, etc.). *Repente*, as a tradition, is a process of stabilizing artifacts in relation to problems and vice versa. What is the goal of this process? What does it achieve? It produces regularities. In our framework, what stands at the center of distributed cognition (delimiting a distributed cognitive system) is not an agent, nor a task, but a rule for producing regularities, which triadically mediates between signs (artifacts) and problems (objects) to produce regular effects (interpretants). This production of regularities can be understood as an inferential activity involving abduction, deduction and induction.

1.1.2.4 Chapter 8: Semiosis as Problem Solving, Problem Solving as Semiosis

Semiosis is understood as a hierarchical, relational process involving artifacts (sign), problems (object), and solutions (interpretants), aligned with Peircean categories. This triadic framework posits problem solving as inherently distributed and dynamic, contrasting with traditional models. The chapter asserts that problem solving involves three hierarchical levels: qualitative framing of problems through artifact qualities (Firstness), reactive manipulation of artifact states to navigate problem spaces (Secondness), and habitualization of solutions for consistent problem-solving under similar conditions (Thirdness). By employing Peirce's ten classes of signs, this chapter further details, from a semiotic perspective, the interaction between artifact, problem, and solution. Each class is associated with specific semiotic operations: framing (rheme), searching (dicisign), and habitualization (argument). These operations are not sequential but reflect a logical hierarchy where higher-order processes (Thirdness) subsume lower-order ones (Firstness, Secondness). The approach emphasizes the irreducibility of the triadic relations in problem-solving tasks, where artifacts embody problems and solutions are mediated through habitual sign-action. This model provides a framework for understanding the distributed, situated nature of cognition, challenging internalist computational theories and extending the scope of problem-solving to encompass a wide range of human activities, including art, literature (and oral poetry) and daily tasks.

1.1.3 Third Part: Meaning and Niche Construction

We describe semiosis as a niche construction process. The focus here is on the emergent self-organization of semiosis in situated environments, and how agents participate in this process (authorship, experimentation in arts).

1.1.3.1 Chapter 9: Meaning Is Niche Construction: Semiotic Artifacts and Cognitive Externalism

Although meaning is context-sensitive and materially extended (situated, embodied, and distributed), its locus is not well-captured by the notion of a physical environment alone. Inspired by biological concepts, we suggest that the locus of meaning is better understood as a niche rather than an environment. Our focus lies on the notion of niches of semiotic artifacts. These niches are structured spaces of fundamental conditions for the stability of semiosis, encompassing factors like situatedness (co-localization) and temporal distribution among communities of agents and their environments. It is only within these niches that artifacts, problems, and rules function as signs, objects, and interpretants. Sign action is niche construction. For instance, consider musical meaning, which relies on artifacts that create a niche for interpretant production, such as metaphors associating music with a moving force, a landscape, or a texture. In this context, musical meaning is external, emerging from situated and embodied artifacts. These external artifacts enable musical pieces to generate regular interpretants that can be communicated in expressions like “the music rises here” or “here comes a smooth passage.” By understanding the concept of a niche, we gain insights into how artifacts, problems, and rules acquire and maintain stable tendencies that define them as components of a sign process.

1.1.3.2 Chapter 10: Artwork Authorship as a Legisign-in-Action

An artifact is commonly understood as an object consciously developed by a specific agent, traditionally seen as its sole author. However, this perspective overlooks the self-organized nature of creativity in niche construction. Instead, we propose a view of authorship as an external sign. In this view, authorship is itself an artifact that plays a crucial role in stabilizing the complex relationship between artifacts and agents within a niche. For instance, authorship provides a means to acknowledge, credit, and attribute the origin of specific artifacts to individual agents. It may also facilitate interpretation by helping an audience to link different media products of the same author and recognize recurring habits throughout them. Authorship can be characterized as a legisign since it regularly associates artifacts (such as artworks) with specific agents (their authors). As a legisign, authorship can be categorized into six distinct classes of signs. This characterization allows us to analyze different aspects of authorship and the various ways it shapes the relationships between agents and artifacts within a niche.

1.1.3.3 Chapter 11: Metasemiotic Experiments

An artwork is an experiment in metasemiosis (semiosis about semiosis). The proposed triadic model of semiosis is utilized as an epistemic tool to describe art experiments. An artistic experiment, as a sign in action, is based on the following principles: (i) the exploration of sign materiality and structure (S-dependent experimental activity); (ii) the exploration of sign referentiality (or referential properties) (S-O-dependent experimental protocols); and (iii) the exploration of interpretative effects (I-dependent experimental protocols). The S-O-I relationship functions as a model capable of distinguishing different modalities of artistic experiments, according to a certain emphasis defined by the terms of the relationship: (i) experiments with the material qualities of the sign; (ii) experiments with the designation or reference relations of the sign, that is, the relations between S and O; and (iii) experiments with the interpretive effects of the sign, that is, with the triad S-O-I. We illustrate this typology with historical cases from the Rupture Group, artists Cildo Meireles, Berna Reale, and Lygia Clark.

Acknowledgments To the researchers, who in different moments, have contributed to the ideas of this book in many discussions and debates, or who have read previous versions of the material: Kobus Marais, Chris Sinha, Frederik Stjernfelt, Irene Mittelberg (*in memorian*), Peer Bundgaard, Lars Elleström (*in memorian*), Ricardo Fusarolli, Ricardo Gudwin, Angelo Loula, Charbel El-Hani, Lúcia Santaella, Winfried Nöth, Kalevi Kull, and Claus Emmeche.

The authors thank, for their support during their field trip in the northeast of Brazil, Ana Fernandes and the singing poets Toínho Batista, Edmilson Silva, Zé Viana, Paulo Martins, Fredi Guimarães, Tindara, Felipe Pereira, Ivanildo Vila Nova, Edmilson Ferreira, Sebastião Dias, Iponax Vila Nova, Antonio Lisboa, Siba, and Guerreira.

Pedro Atã thanks the Faculty of Humanities of the University of the Free State for his Postdoctoral Research Fellowship.

João Queiroz thanks CNPq (National Council for Scientific and Technological Development) for the support through the research productivity scholarship (PQ2 308355/2023-7) and the regular research project (Emerging Groups; 404770/2023-1).

We thank Berna Reale, and Fundação Lygia Clark for generously permitting us to use images of their artworks.

Chapter 2
Active Semiotic Externalism

2.1 Cognition Is Made of Relations

What is cognition made of, and how does it work? In Peirce's cognitive semiotics,[1] cognition is made of sign action, or semiosis, an embodied, situated and distributed process that integrates representation/inference, dynamic situated relations, manipulation of artifacts, and culture/communication. We call this view active semiotic externalism.

In the context of theory of relations, semiosis is modeled as a triadic relation, including its degenerate variations.[2] To conceive of cognition as irreducibly semiotic involves some degree of commitment to the idea that relations are explanatorily more fundamental than individual entities. Peirce can be considered a metaphysical foundationalist and realist about relations (Burch 1997, p. ix). Peirce's pragmatism, a philosophy of action *par excellence*, doesn't answer the question "What is cognition made of?" with any kind of entity, such as thoughts, concepts, representations, or mental contents, but rather with a kind of active relation: sign-action or semiosis.

---

[1] This description primarily applies to Peirce's late, post-1903, philosophical system (see Queiroz and Stjernfelt 2019).

[2] Peirce refers to a degenerate triad as a "mere juxtaposition" (CP 1.371), a "mere combination" (CP 1.363), of "congeners" (MS 717). The sign is an example of a triadic relation. A sign relation, S-O-I, involves a dyadic relation, S-O. If the triadic relation can be without the third term (I), then it is a degenerate relation in the first degree, and it is an index. If it is possible for it to be without the second and third terms, then it is doubly degenerate, and it is an icon. In the essay, which, according to Fisch (1986, p. 35), is one of the best ever written about the theory of the sign, "New Elements" (EP 2: 300, NEM 4: 238–263, 1904), symbols are defined as genuine signs, indices as signs degenerated in the first degree, and icons as signs degenerated in the second degree, or doubly degenerated (NEM 4: 241–243).

P. Atã, J. Queiroz, *Mind as an External Sign: A Semiotic Externalist Philosophy of Cognitive Science*, Synthese Library 531,
https://doi.org/10.1007/978-3-032-08988-5_2

At the foundation of Peirce's semiotics is the study of relations and how some relational forms involve others.[3] This systematic understanding of relations is treated, in Peirce's philosophy, as a system of categories.[4] Peirce's categories were influenced by a long tradition of thinking in categories, from Aristotle to Kant and Hegel (CP 8.329).[5] Each category corresponds to an irreducible logical, phenomenological and metaphysical relational form. Peirce calls them Firstness, Secondness, and Thirdness. Each category is thus both a fundamental and irreducible relational form, and a fundamental and irreducible element of experience. This mapping between relational forms and elements that constitute a mind is a central methodological innovation behind Peirce's philosophy. It offers a method to treat mental phenomena and sign action in a systematic, quasi-formal, way, and it differentiates the Peircean approach from other phenomenological approaches to cognition.[6]

In logic, the categories answer the problem of how to reduce all varieties of complex relations to a short list: monadic, dyadic and triadic relations (Parker 1998, p. 43; Houser 1997, p. 14; Murphey 1993). In phenomenology, they answer the problem of what are the most fundamental components of mind/mental experience (Atkins 2018).[7] First, mere feelings, which are internal relations and states of experience, and which only potentially tells us of reality. Second, dynamic interactions, or the relation between experience and something external to it and which constrains it and compels it, and which tells us that there are actual facts in reality. Third, mediations, which allow experience to operate vicariously, so that one can experience a part of reality through another part, and which tells us that there are

[3] Peirce is not only interested in a theory of cognition based on a general theory of relations that functions as a "stecheology" of thought (CP 1.191), but at the core of his philosophical system is the concept of relation (MS 931: 396; MS 931: 396).

[4] From the 1870s–1880s onwards, the categories come to describe irreducible types of relation (monadic, dyadic, and triadic), each serving a distinct and necessary unifying function by which objects are made intelligible (see Hausman 1993, p. 109). Instead of deriving the categories from the analysis of sign relations, as he had done in 'On a new list of categories' (W 2: 49), Peirce presents the categories directly as three classes of relation—monadic, dyadic, and triadic ('One, Two, Three: Fundamental Categories of Thought and of Nature,' W 5: 242). This approach offers the advantage of generality since all possible logical relations, including the sign relation, fall within one of these three classes.

[5] The initial exposition of the categories is Kantian, a philosopher whom Peirce identified with for a substantial period (CP 4.2), and from whom "On a new list" is a "reconstruction" (CP 1.563, 4.2, 8.329). Primarily, the categories are defined as "conceptions that bring the variety of impressions down to a unity," specifically the unity within a proposition. Additionally, they echo Kant's definition of number—"essentially the unity resulting from the synthesis of the variety in a general homogeneous intuition" (Savan 1952, pp. 186–187).

[6] Perhaps we should distinguish cognition from phaneron. What is the relationship between the two? On the one hand, not all phaneron is 'cognition'; for example, feeling and struggle. Only the thirdness of the phaneron is cognition. On the other hand, perhaps not all cognition is a 'phaneroscopic' experience or involves experience.

[7] According to Pape (1997), the categories are a metatheoretical principle whose development required formal analyses, phenomenological investigation, and metaphysical speculation. The construction of this principle, whose mathematical foundations connect different methods, associates various domains, both formal and empirical.

general regularities in reality. In metaphysics, the categories are processual rudiments of the world, that is, they express how reality is in process. In semiotics, this short list describes signs, objects, interpretants, the irreducible and conceivable patterns of semiotic processes (classes of signs) and their role in cognition and communication. From the three fundamental categories, more extensive classifications of signs are derived, such as the famous trichotomy icon—index—symbol, and the later and more comprehensive 10 classes of signs, which we will explore later.[8]

The system of categories describes cognition as a triadic relation, semiosis, which is also a hierarchical process internally composed of relations. Firstly, a sign has internal relations, or relations to itself, which express its intrinsic properties, such as its possible states and transitions between these states. For example, phonology, morphology and syntax are internal relations that define a space of possibilities for language. Secondly, the sign reacts with objects in a situated environment, so that its space of possible internal properties is constrained by this environment. For example, although grammar allows, equally, the occurrence of the propositions "the dog is flying" and "the dog is running", the observed behavior of the domestic animal, co-situated with language, constrains language unequally between these two propositions. Thirdly, signs produces more signs (called interpretants), in such a way that the sign-object relation itself is dependent on this further production of signs, and is thus distributed in time and space across several instances of sign-action. For example, "the dog is flying" is a proposition that someone utters so that someone else (or him/herself in the future) may interpret it, with this interpretant being itself a sign which reinstantiates the former. This chain of signs producing new signs mediates, in a distributed fashion, all instances of the word dog that a person has ever encountered in his/her life, and allows him/her to interpret "the dog is flying" as a fantastical proposition, because he/she has a general understanding of "dog" which grew throughout these many instantiations. For Peirce, the third relation involves the other two, and semiosis can only occur if all three relations are involved.

The triadic system of categories is Peirce's ultimate thinking tool[9]—it encapsulates, in a bite-sized model, the central theses of Peircean cognitive semiotics, including cognitive externalism, and it leads to a way of conceptualizing and analyzing phenomena as triadic processes, avoiding dualisms and reductionisms. The categories and the classes of signs that they produce can also be seen as an analytic tool: they allow one to identify and classify the components of different semiotic phenomena, to group together phenomena of the same class and to understand how these different groups are related. Finally, they also work as a high-level conceptual

[8] The classifications were developed at various moments, from 1867 to 1908, and are organized into what can be considered systems of classes: 3 classes, 10 classes, 28 classes, and 66 classes of signs. See: Jappy (2018, 2021), Farias and Queiroz (2017a, 2006).

[9] Categories are 'best viewed as recursive heuristic conceptions. To describe them as heuristic implies that they are first and foremost guides and goads to inquiry…. [They] are not procrustean beds; they are not a priori molds into which the facts must be poured. They are relevant guidelines for … fruitful [interpretation]' (Colapietro 1995, pp. 29, 30).

model of how semiosis/cognition works, by providing a systematic map of conceivable states that a semiotic process can be in, a structure of dependence and presuppositions among those states, and the transitions between them.

2.2 How Does a Process Philosopher Explain Cognition?

We often describe mental occurrences as if they were passive objects manipulated by an individual—someone generates an idea, maintains a belief, conveys a thought, processes information and so forth. This manner of expression assumes a dualism between a conscious actor who intends and takes actions and a form of content or representation that becomes the recipient of these actions. It also portrays cognition as if it were composed of concrete, substance-like information entities that can be subjected to manipulation. This illustrates a perspective rooted in a substance-oriented ontology. Conversely, we have the option to perceive mental phenomena as actions. For instance, in the framework of pragmatism, beliefs are not abstract concepts stored in a representational system, but rather dispositions to engage in specific activities.

It matters whether we treat mental phenomena as things or as processes because things and processes have different properties. A thing is defined by an unchanging substrate. A process is defined as a certain pattern of activity. A thing is particular, and we can think of classes of things as groups of particulars. A process can be a general pattern distributed throughout many particular instances, thus defying the notion of class as a collection of individuals. The nature of a thing is to remain unchanged, with any alterations considered exceptions explained by unique circumstances. Conversely, the nature of a process is to continuously transform, evolve, or grow. A state of unchanging stability in a process is considered exceptional and should be accounted for by specific circumstances.

A substance, as defined in its Latin origin as "something that stands under or grounds things" (Robinson 2020a), is a stable and internally undifferentiated particular. Substance ontology regards properties as intrinsically possessed by substances, and views change and transformation as arising from, and consequently secondary to, particulars and their inherent properties. Thinkers aligned with substance ontology tend to build models and theories based on compositions of particular and unchangeable units, simple and internally undifferentiated (Seibt 2018), and they tend to pursue explanations of their subject matters in terms of the inherent properties of those units. The focus is on "things, attributes, and timeless relations" (Rescher 2000, p. 33). That is, substance thinkers try to boil down processes and organizations to constitutive units that are not in process and not organized, but which have properties or attributes that can explain how the processes and organizations come to be constituted. In this philosophical framework, change, transformation, and appearance of novelty need to be explained in terms of the properties of substantial units. However, as these explanatory building blocks are simple and internally undifferentiated, they themselves do not change or develop—any change is explained in terms of the properties of more fundamental building blocks, ultimately tracing back to fundamental point particles (Bickhard 2011; Campbell and Bickhard 2011). Without some external condition for change, explanatory building blocks remain static. Change is regarded as secondary and external to identities. Thus, substance thinkers feel the need to explain why substances undergo change, assuming that the natural course for something is not to change but to remain as they are. Nicholas Rescher sums up three fundamental tenets of substance ontology's "revolt against process":

1. The appropriate paradigm for ontological discussion is a thing (most properly a physical object) that exhibits qualities (most properly of a timeless—i.e., either atemporally or temporally fixed—character).
2. Even persons and agents (i.e., "things" capable of action) are secondary and ontologically posterior to proper (i.e., inert or inertly regarded) things.
3. Change, process, and perhaps even time itself are consequently to be downgraded in ontological considerations to the point where their unimportance is so blatant that such subordination hardly warrants explicit defense. They may, without gross impropriety, be given short shrift in or even omitted from ontological discussions. (Rescher 2000, p. 34)

In process metaphysics, a process is defined as "coordinated occurrences of change in reality, an organized family of occurrences that are systematically linked to one another either causally or functionally" (Rescher 1996, p. 38). Within process ontology, development, transformation, and change are seen as intrinsic aspects of reality, while stable particulars are considered as outcomes derived from and secondary to these ongoing processes. In this framework, properties and attributes tend to be seen as relational and emergent, rather than inherent to substances (Bickhard 2011).

Process philosophers prioritize a state of continuous transformation over seemingly static building blocks, viewing their research objects as evolving organizations that arise, develop, and eventually dissipate within this unceasing flux. As stated by Dupré and Nicholson (2018, p. 13), "What we identify as things are no more than transient patterns of stability in the surrounding flux, temporary eddies in the continuous flow of process." Consequently, stability and the concept of "thingyness" must be explained in terms of more fundamental events and transformations, leading process thinkers to seek explanations for why things, in some aspects, remain as they are, acknowledging that the natural tendency is toward change.

Two central concepts in process philosophy are dynamicity and emergence. Dynamicity, as described by Dupré and Nicholson (2018), pertains to continuous temporal change. As articulated by Johanna Seibt (2018, par. 1), process philosophers posit that "the dynamic nature of being should be the primary focus of any comprehensive philosophical account of reality and our place within it." For them, it's inconceivable to envision entities apart from the arrow of time; instead, identities are an integral part of the universal process of flow and cannot be considered in isolation. The emphasis isn't to diminish stable identities but to reframe how one explains both stability and change. Process thinkers distinguish between varying rates, conditions, and regularities of change, with change being fundamentally explanatory. From this perspective, stability itself is a type of flow, where stable structures undergo regular transformations in time, albeit at different rates and under different conditions compared to neighboring structures. Stable organizations exist within and through transformational flows. Accordingly, what is more important for process thinkers is not to explain what motivates a structure to change—it is assumed that every structure is subject to change. It is more important to explain why and how a flow gets stabilized and how this stability forms conditions for the development towards another form of stability and/or towards eventual dissolution or dissipation. In other words, it is more necessary to explain stability in change than change per se.

Emergentism is an explanatory framework that tries to account for true novelty within a naturalistic view of the universe. Emergence is an asymmetrical or non-linear relation between levels and/or domains of a phenomenon. While it can be explored from a substantialist perspective, it finds a more plausible home within a process-oriented ontology (Bickhard 2003; Campbell and Bickhard 2011). A substance-oriented understanding of emergence has the challenge of explaining how physical matter (substances) acquires causal powers due to non-physical (organizational) processes (see Kim 1999). For example, the substance-emergentist has to explain how mental activity supervenes, ultimately, on the properties of physical particles. Differently, a process-oriented understanding of emergentism emphasizes that causal powers are held by organized processes. Any kind of stable substance (including matter) is a result of organization (e.g., matter is a result of interactions in energy fields, see Bickhard 2011), so that emergence is a natural consequence of the way the universe works: different organizations are capable of generating different causal powers. Consequently, emergence becomes a natural consequence of the universe's functioning, where diverse organizations generate distinct causal powers. A process-oriented emergentist doesn't need to explain how cognition supervenes on physical particles but rather how cognitive relationships can occur among non-mental structures. A process-oriented understanding of emergence can be characterized as a doctrine of the primacy of interlevel relations and organizations over substances and units (see Campbell and Bickhard 2011).

2.3 Categories: A Tool for Investigating Relations

Peirce is a process philosopher (Lorino 2014; Queiroz and El-Hani 2006a; Hausman 2002; Rescher 1996) whose epistemology and metaphysics, particularly through the principles of pragmaticism (Legg and Hookway 2019), synechism (Esposito 2005), and tychism (Bergman and Paavola 2020), regard the fundamentals of meaning and reality as open-ended dynamic processes. The orientation towards processes in Peirce's cognitive semiotics is patent in how the problem of the nature of cognition is framed as a problem about relations. To ask "What is cognition constituted of?" is equivalent to asking "What kinds of relations constitute cognition?" The first step to address this problem is to clarify what kinds of relations are there to start with. We have to start from a list of relations in order to select and justify which kind of them better explains cognition. Thus, at the foundation of Peirce's semiotics is a logical investigation of relations. Peirce's insight was that the logical composition of concepts, in cognition, was akin to a bonding[10] of relations with other relations, in which simple relations participate in more complex relational structures.

Relations can be classified in terms of how many bonds or connections they are able to form: a monadic relation makes only one connection, a dyadic relation makes two connections, a triadic relation makes three connections, a tetradic relation makes four, and so on. This gives us an infinite list of relational forms. However, according to Peirce's "theorem of irreducibility of relations" (Herzberger 1981; Burch 1997; Houser 1997, p. 14), "proof of irreducibility" (Ketner 1986, pp. 376–377), or "thesis of irreducibility" (Burch 1991, 1997), we don't need such an infinite list: monadic, dyadic, and triadic relations are necessary and sufficient as a list.[11] There are two clauses involved in this thesis: first, that at least some triadic relations are irreducible to combinations of dyadic and monadic ones—thus, it is necessary to include triadic relations on the list; second, that all tetradic and higher adicity relations can be reduced to combinations of triadic relations (Koshkin 2022),—thus, the short list of monadic, dyadic, and triadic relations is already sufficient.

An intuitive way to grasp Peirce's irreducibility thesis—which is by no means a proof or demonstration of it—is to consider how any form of combination is itself a triadic relation. A combination of A and B is a triadic relation involving A, B, and the combination itself of them. Thus, at the core of any n-adic complex of relations is the operation of combining together relations, or building relations of relations, and this core operation is itself a triadic relation. As Peirce puts it: "nothing can be built out of other things without combining those other things, and combination is itself manifestly a triad" (EP 2.366). In Peirce's cognitive semiotics, this insight is translated into the propositions that all cognitive processes are triadic relations, in the sense that they are, in essence, a processual combination of relations, or relations of relations. Each relational form is associated to ontological properties (modes of being) and logical modalities, so as to form a philosophical system of categories, which is the foundation for Peirce's semiotics and for active semiotic externalism: Firstness, Secondness, and Thirdness. In brief, the categories can be defined as follows:

---

[10] The notion of "bonding" comes from an analogy, made by Peirce, between the logical composition of concepts and the physical composition of chemical ions (Burch 1991), with the important difference being that such a logical bonding was not between atomic units, but between relations, that is, relations between relations.

[11] The status of Peirce's reduction thesis has been controversial, especially as results in logic in the twentieth century indicate that all relations can be reduced to dyadic relations (e.g., Quine 1995). Burch (1991) has finally provided an exposition and proof of Peirce's irreducibility thesis. In his argument, the crucial aspect is a different understanding of "reduction" in Peirce's thesis, when compared to the theses that reduce relations to dyads. While reduction to dyadic relations can be achieved with the use of some logical systems, it is precluded by the use of others. This approach allows the proof to be derived, but it is also open to the accusation that the tools used to derive the proof were specifically designed to make it work. Koshkin (2022) has recently proposed a solution to avoid such an accusation.

Firstness, the category of relations that can only make one connection, is the mode of being in which something is as it positively is, without regard or relation to anything else (MS 460: 5–21, 1903; MS 575, 1886; CP 5.299). It can be characterized as lacking determination (cf. also MS 277, 1908). Therefore, Firstness as a mode of being associated with the modality of possibility. It is the category of vagueness and novelty—"the mode of being which consists in its subject's being positively such as it is regardless of anything else. That can only be a possibility" (CP 1.25). It is the "being of a feeling, in itself, unattached to any subject", a mere "possibility floating in vacuo" (CP 6.342).

Secondness, the category of relations that can make two connections, is the mode of being "which is as it is relative to a Second but regardless of any Third." It is a kind of reaction (see CP 6.200), or "arbitrary brute-force action upon other thing" (CP 6.342). Secondness can be associated with the modality of actuality (Parker 1998; CP 6.455). The actuality of a thing is simply its occurrence. Actuality is the realization of a possibility, without thereby making reference to any general law or interpretation. Peirce considered "the idea of any dyadic relation not involving any third as an idea of secondness" (CP 8.330).

Thirdness, the category of relations that make three connections, and thus of relations that can combine other relations, is the category of mediation, habit, generality, and conceptualization (CP 1.340). The example par excellence is sign action (semiosis) in which a sign is related to an object through the mediation of an interpretant. It can be associated with the modality of necessity, in the sense of rational necessity and necessitation (CP 6.342), since a triadic relation is one in which an occurrence is connected to a reason, that is, a general pattern of occurrences or a habit. This is also the category of semiosis in general.

The system of categories is used by Peirce to define and flesh out a model of cognition. As we will explore in more detail in the next sections, there are at least three ways in which the categories are applied in this model. Firstly, the hierarchical ordering of Firstness, Secondness, and Thirdness is used to define the nature of cognition as a process of Thirdness, and thus distinct from relations which are merely reactive (Secondnesses) or merely qualitative (Firstnesses). Secondly, considering Thirdness, or triadic relation, the hierarchical ordering of categories helps to define the three components of the triadic relation, as ordered terms: a triadic relation includes a First term, a Second term, and a Third term. The order matters here, because it defines specific roles for each component of the relation: the First of the triadic relation of cognition corresponds to the role of the sign, the Second to the object, and the Third to the interpretant. Finally, the notion of hierarchical ordering of categories is applied one more time to explore the relational nature of each of these terms. That is, the sign, the object, and the interpretant, may each of them be a monadic relation, a dyadic relation or a triadic relation. This is the basis for the notion of classes of signs. In summary, the triadic system of categories enables the conceptualization of cognition as a triadic relation which combines other relations. This involves ordering the component relations as sign, object, and interpretant.

Additionally, each component relation is categorized as either a monad, a dyad, or a triad, leading to the derivation of a typology of sign classes.

We may accept Peirce's notion of irreducible categories in a weaker or stronger sense. In a weaker sense, the categories are a modeling framework dependent on a certain logical apparatus (Burch 1991), and which is especially appropriate to understand complex processes. They are phenomenological in scope, not metaphysical, although they embody the ontological proposition of a universe constituted by processes and which possesses properties such as chance, habit, and emergence. In a stronger sense, the categories are metaphysical rudiments of a universe foundationally composed of relations, and in which the most fundamental kind of relation, from which the others stem, are triadic relations.

2.4 The Phaneron and Its Elements

Equipped with this short list of necessary and sufficient relations, we can apply them to mental phenomena. Peirce introduced the word *phaneron* to refer to the "collective total of all that is in any way or in any sense present to the mind, quite regardless of whether it corresponds to any real thing or not" (CP 1.284). The phaneron ranges from the faintest feelings to the most explicit and self-reflexive reasonings. Following Peirce, we can analyze the *phaneron* in terms of the relational forms present in the previously constructed short list. This was Peirce's version of phenomenology, which he called *phaneroscopy*.

Peirce understood phaneroscopy to provide a context of empirical justification for the three categories, in the sense that phaneroscopic experiments are a direct practical application of the categories which, up to this point, only exist as logical-mathematical entities. In the path towards a theory of cognition, phaneroscopy corresponds to a first step of close inspection and examination of the phenomenon the theory is supposed to explain. The goal of phaneroscopy is to examine the mind as it appears phenomenically, and uncover the fundamental elements of mental experience. Phaneroscopy examines the phaneron itself, its "formal elements" (CP 1.284). Its main questions are: What are the indecomposable elements of the phaneron? (MS 292: 71–75) How do they interconnect? The result is a catalog of the "departments of mental action" (CP 7.539). The main question, in a more simplified form, can be put as follows: can the extraordinary diversity of events and mental states exposed to observation be classified in a short list of elements? An element, here, is a relational form, not a kind:

> We are to consider what forms are possible, rather than what kinds are possible, because it is universally admitted, in all sorts of inquiries, that the most important divisions are divisions according to form, and not according to qualities of matter, in case division according to form is possible at all. (Peirce 1998, p. 362)

The premise involved in this investigation is that the phaneron is relational in nature, so that phaneroscopic elements correspond to relational forms. Phaneroscopy involves a method for separating relations from relations. The short list of three categories of relations assumes a hierarchy of dependence between relations, so that there are different degrees of separation between them. Although different kinds of relations can be distinguished from one another, it is also the case that simpler relations are part of more complex relations. Categories are separated in some sense, but are together in another. To account for these hierarchical relations, Peirce differentiates between degrees of separation: dissociation, prescission, and discrimination (Atkins 2018). Dissociation is the stronger form of separation. We dissociate two ideas when they are "so little allied that one of them may be present to the consciousness in an image which does not contain the other at all" (W5: 238). For example, we can dissociate red (as a phenomenon) from blue, because they may manifest in experience completely independent from each other.

Prescission is a separation by concentrating on one element and neglecting all others. It can be likened to abstraction. For example, upon hearing a piece of music, we can prescind (abstract) the melody from the sound that conveys it, even though we cannot dissociate one from another, that is, the melody does not manifest in experience independently from sound. We perform the prescission by concentrating on the melody (pitch variation) while neglecting other elements that compose a sound, such as timbre. In this case, melody and sound are together in the sense that they cannot be dissociated, but are separate in the sense that one can be prescinded from the other. Note that if we want to separate sound from melody, instead of the opposite, we are able to perform a dissociation, since there are plenty of occurrences of sounds which are not melodic. Another example is the prescission of spatial extension from color. Upon seeing a field of red surrounded by blue borders, we can prescind the spatial extension of the color from the color itself, by focusing on its shape and size and neglecting its coloration, in a similar way that we can prescind melody from sound. Prescission is capable of establishing hierarchical ordering in experience, in the sense that some elements may be precinded from others, without the opposite being true.

Discrimination, or distinction, is the weakest form of separation. Here, it is enough that we represent two parts as separate, regardless of being able to abstract them in experience (prescission) or to actually experience them separately (dissociation). For example, while, upon seeing a field of red, we can prescind the spatial extension of the field from its coloration, we cannot prescind the color red from a spatial extension (this example is adapted from Aames 2021). Any experience, whether actual or imagined, of the color red, will have to be manifest together with a spatial extension to it. However, this does not prevent us from being able to distinguish the concept itself of the color red by using the expression "the concept of red without spatial extension", which we can manipulate within a language to write about red as a concept independent from the concept of spatial extension, just like we are doing right now in this paragraph, and regardless of being able or not to form an imagetic experience of the concept itself of red without spatial extension.

In phaneroscopy, we perform dissociations, prescissions, and distinctions between parts of experience. By performing dissociations, we arrive at experiences that manifest as separate from other experiences. For example, the simple experience of red, or of blue. They are relations, since experiences themselves are relations, but disconnected from other relations. Thus, they are the phaneroscopic correspondent to monadic relations: simple qualities or predicates. When we are unable to perform dissociations but can perform prescissions, we arrive at experiential relations that are correlated to other relations. For example, the acoustic experience of a melody is correlated with a sound, and the visual experience of red is correlated with a spatial extension. They are relations that involve relations (melody involves sound; red involves spatial extension), the phaneroscopic correspondent to dyadic relations. Finally, when we are unable to perform dissociations or prescissions, but can perform distinctions, we arrive at experiential relations of representation. For example, we can separate the color red from any spatial extension by resorting to an abstract concept of the color red expressed in a representation, such as the linguistic representation "the concept of red regardless of spatial extension". This is an example of a triadic relation, which involves dyadic relations, which in turn involves monadic relation (the linguistic representation of the concept of red involves the relation between space and color so that space is separated from color). Thus, a monadic relation can be dissociated from any other. A dyadic relation, however, involves a monadic relation which can be prescinded from it, but not dissociated. Finally, a triadic relation involves a dyadic relation and a monadic relation, which can be distinguished from it, but not prescinded or dissociated.

Through the performance of the different degrees of separation within the phaneron, the short list of relational forms gives rise to a short list of departments of mental experiences: feeling (*primisense*), which corresponds to monadic experience, or experience that can be dissociated from others; *altersense* (sense of otherness), which corresponds to dyadic experience, or experience which can prescinded from others but not dissociated; and *medisense* (sense of mediation or representation) which corresponds to triadic experience, or experience which can be distinguished, but not prescinded nor dissociated from others. According to this division, three fundamental types of experience must be explained. (i) "the momentarily present contents of consciousness" (CP 7.551) (e.g., the distinct experience of red), (ii) the experience "of a directly present other or second, withstanding us" (e.g., the experience that the feeling of red requires spatial extension), and (iii) the experience of synthesis or mediation (see also CP 1.378; CP 2.85; CP 1.306; EP 2:150; CP 1.324; CP 1.332; CP 7.544) (e.g., the experience that red can be mediated by the general concept of red, which distinguishes red from the spatial extension of red).

The collective total of experience, the phaneron, is formally composed, first, of vague dissociated feelings present to a consciousness and regardless of anything else (e.g., a shape, color and movement is present to my mind); second, of the dyadic experience that these feelings are connected to some otherness "out there", independent from one's pure will (e.g., this particular shape, color and movement are connected in some way, independently of my conscious will, and thus constitute something "out there" that I cannot imagine away); and, third, of the experience that there is mediating element that synthesizes the otherness and one's experience of it (e.g., the concept of a "frog leaping" mediates between my ability to experience a pattern of color, shape and movement, and my inability to imagine them away or to separate them, that is, I cannot imagine them away because, as I recognize, they are the shape, color and movement of a frog leaping, and not of my own imagination).

We can characterize monadic experiences merely as presence to consciousness. But when we describe altersense, dyadic experience, we are forced to further qualify a presence to consciousness: is it an action exerted by conscious will or against conscious will? Dyadic experience is the experience of something other, which acts on one's conscious experience, regardless of one's conscious will. Altersense can be summed up as the experience of being acted upon. But one can only experience being acted upon if there is some difference, or opposition, between what one expects is going to happen and the actual outcome. Peirce gives simple examples based on pushing and touching: you press against a door, the door resists against your force; that experience of acting and being resisted against is precisely the experience of the door as something other than your conscious will. Altersense is described as a dyadic pair of action/resistance to action, or expectation/surprise. Altersense is the phenomenological basis upon which we experience facts of a reality independent of our mental lives. Finally, medisense is the phenomenological basis upon which we can learn about this reality. It corresponds to the sense that the phaneron is rational, governed by regular patterns. The term "medisense" alludes to the notion that experience is not only formed by direct relations to the here-and-now, but is mediated by patterns that carry over from past occurrences and towards future (anticipated) occurrences. There is a hierarchy involved in this sequence of three categories. Dyadic experience has to involve monadic (a sensation that is compelled by something Other); Triadic experience has to involve dyadic and monadic (a representation that mediates between a feeling and its correlated Other).

2.5 The Semiotic Triad

Although the phaneron is composed of monadic, dyadic, and triadic relations, in Peirce's theory only the latter are sufficient and necessary to describe genuine cognitive/semiotic operations. Thirdness is the phaneroscopic domain of semiosis, the part of the phaneron that can be recognized as meaning, cognition, communication, interpretation, thought, and so on, as opposed to merely dyadic reactions or monadic feelings. An important difference between semiotic and non-semiotic processes is that non-semiotic processes can be decomposed into dyadic or reactive processes, while semiotic processes are irreducibly triadic. Semiosis is a concept that describes the most fundamental relations involved in processes of cognition (EP 2:646).

Semiosis is a triadic logical relation, that is, a relation that irreducibly involves three sides: a first element, which has qualities in itself (monadic, or first relation), but which stands to a second element (dyadic, or relation of a relation), only so far as the dyadic connection between them is mediated by a third element (triadic relation). Peirce calls the first element of the triad sign (S), the second object (O), and the third interpretant (I). This is a recursive application of the categories: a semiotic relation is a phaneroscopic thirdness (S-O-I triad), and as such it involves a

Secondness (S-O dyad) and a Firstness (S monad). sign, object and interpretant are irreducible terms of a semiotic relation (CP 5.484, EP 2:171).

Triadic irreducibility is a requirement of any process that we might regard as "cognitive", "representational", or related to "meaning"—"A sign, or representamen, is something which stands to somebody for something in some respect or capacity" (CP 2.228). The expression "stand for" is metaphorical (de Waal 2013, p. 79). This "standing for" relation involves a constraining or determination (NEM 4:54). Therefore, if we consider only dyadic relations, S-I, S-O, or I-O, or the elements in isolation, we cannot deduce how they would behave in a triadic relation, S-O-I (EP 2:391). Irreducibility of semiosis can be understood in terms of the non-deducibility of the behavior of the logical-functional elements of a triad on the grounds of their behavior in simpler relations (Queiroz and El-Hani 2006a).

In principle, anything can take the role of sign, object, or interpretant (MS 634:18). They are not necessarily mental representations or models, abstract concepts, intentional contents, physical entities, etc. What is decisive for defining S, O, and I is the role they play in a triadic relation, and not whether they are internal or external to the brain and the nervous system. The role of S can be identified only in the mediative relation that it establishes between O and I. Similarly, the role of O is identified in the relation by which it determines I through the mediation of S. And, finally, the role of I is identified by the fact that it is determined by O through S. Nevertheless, by virtue of their role in the triadic relation, sign, object and interpretant are associated to distinct logical and phenomenological characteristics.

The Firstness of the triadic relation is the sign. As a first, it is associated with what is immediately present to experience, as well as with vagueness and possibility. The sign is sometimes referred to as 'representamen' or yet 'sign-vehicle'. Throughout this book, we will also use the term "artifact" or "semiotic artifact" to refer to it, especially when we treat semiosis as problem solving (Part 2). The sign, or artifact, can be a process of any kind: words on a page, assertions in an internal dialogue, visual images of all kinds, whether natural (e.g., smoke as a sign of fire) or produced (e.g., photographic images, painterly images, mental images manipulated with one's eyes closed…), sounds, diagrams of all types (such as mathematical formulas, geometric diagrams, etc.) and all sorts of media (music, film, print media, digital media….). signs always have a material form of some kind, that is, they are not in any sense ideal forms, and must determine physical effects, however complex this determination may be.

The semiotic artifact, in itself, is associated with vagueness and possibility because, if we prescind (isolate) the logical properties of the sign in itself from the rest of the triadic relation, we are left with a mere possibility for producing meaning. Any sign has the potential to stand for a multitude of objects, in many possible ways. For example, a smoke may be a sign of fire burning, but it can also be a sign of the direction and intensity of the wind that blows the smoke. When we consider merely the smoke in itself, we can merely conclude a space of possibilities for generating interpretative effects, based on the material properties of the smoke and how it *may* be coupled to a situated environment. Thus, the semiotic artifact in itself is vague.

The secondness of the triadic relation is the object (or, more precisely the sign-object relation). As a second, it is associated, in phenomenology, with brute force reaction and compulsion by an otherness. In logic, it is associated with actuality and the particularity of events and occurrences. In semiosis, the sign, or representamen, has to be correlated to something else, which it conveys or "represents" somehow. The object is something other than the semiotic artifact, and which acts as a decisive factor in how the artifact affects an interpreter. An object can be anything conceivable or communicable: "By an object, I mean anything that we can think, i.e. anything we can talk about" (MS [R] 966).

Semiosis is a mediated production of effects by an object on an interpretant through a sign. The word "Rain" may be a sign of an approaching storm, but it may also be a metaphor for some emotional state, in a song or poem, for example. In this case, the approaching storm, in the first case, or the emotional state, in the second, is a factor constraining how the sign is uttered and how it affects an interpreter. The sign-object relation is an actual relation. The object is the "progenitor" (MS [R] 499), or "quasi-agent" (MS [R] S104) that determines the sign, regardless of whether it exists in reality or not. That is, whether the object is a real process or a fictional idea, that object is somehow, by virtue of a complex chain of actual determinations, the ultimate originator of the sign.

The thirdness of the triadic relation is the interpretant (or, more precisely, the fully triadic sign-object-interpretant relation). As a third, it is associated with mediation and representation in phenomenology, with general necessity in logic, and with regular tendency in ontology. The interpretant is not to be confused with interpreter. An interpreter may be a person, a non-human animal, a macromolecule, a machine, or yet a combination of them, homogeneous or heterogeneous (Queiroz et al. 2011). An interpretant, on the other hand, is an effect produced, by the sign-object relation, in that interpreter. Interpretants range from logical conclusions (such as understanding an argument) to simple behavior (such as turning one's head to the direction indicated by a pointed finger). It is associated with necessity in the sense that it realizes a regularity, or a habit, which is being conveyed from an object, through a semiotic artifact. If, upon seeing a pointed finger, a person turns his/her head to the direction indicated, he/she has realized and confirmed that there is indeed a representation, or a semiotic mediation, between the pointed finger and the direction of the object pointed. Without the moving head, we would have only a certain object that determines a finger to be pointed, but no real confirmation that the finger is a pointing gesture with the purpose and function of indicating the object. The interpretant realizes and confirms a regular and general connection between object and sign. It is the element that renders this connection (or, confirms it as) logically necessary.

Semiosis can also be understood as a process of self-regulation, that is, an action of self-organized regulatory tendency which begets more of itself. Peirce's semiotics has shifted from an early notion of meaning as a qualitative property mediated by a sign, to a relational notion of meaning as a "law relating operations performed upon the object or conditions of perceptions to perceived effects" (Flower and Murphey 1977, p. 589). The idea that processes of meaning are self-regulatory processes is expressed through the concept of habit. For Peirce, "what a thing means is merely what habits it involves" (CP 5.400). (A thing is also defined in terms of habits—"What we call a Thing is a cluster or habit of reactions, or, to use a more familiar phrase, is a centre of forces"; CP 4.157.) A habit is a rule of action (CP 5.397), a tendency or disposition to act in certain ways under certain circumstances (see CP 5.480), or, simply, the "permanence of some relation" (CP 1.415). The definition of habit as "permanence of some relation" characterizes habit as a relation-in-action.

Semiotic processes are regularities of temporal/material events and their properties. A regularity of material events and their properties is a generalization of relations between different kinds of materialities. A regularity of temporal events and properties is a generalization between past, present and an anticipated future. Semiosis is the development of some regularity of relations throughout various different semiotic artifacts, but according to an actively developing pattern, a habit. The conception of meaning as habit is tied to pragmatism (CP 5.491; Murphey 1993, p. 315; Bergman 2016).

Pragmatism was originally formulated by Peirce as a theory of meaning, a methodological principle or rule to clarify ideas, concepts and propositions (Hookway 2002; Legg and Hookway 2019). This principle states that the meaning of a sign implies likely events to occur in a course of experience, according to a set of antecedent conditions (CP 5.468). The meaning of a sign is not a given content and is never experienced as an essential form, but only as conditional rules of action that generate effects under certain circumstances. As circumstances vary, signs take on new habits. In this formulation, semiosis is an open-ended process, constrained by rules and likelihoods, but open to novelty.

Finally, semiosis can also be understood in a communicational sense, as a behavioral pattern that emerges through the intra/inter-cooperation between agents in a communication act, which concerns an utterer, a sign, and an interpreter. Semiosis and communication processes are defined in terms of the same "basic theoretical relationships" (Ransdell 1977, Pietarinen 2006), i.e., in terms of a self-corrective process whose structure exhibits an irreducible relation between three elements. In a communication process, "[i]t is convenient to speak as if the sign originated with an utterer and determined its interpretant in the mind of an interpreter" (MS 318: 11), and the interpreter may become an utterer in a subsequent communication process, trying conveying the same meaning embodied in the sign, thus establishing a chain of communicative events. One of the definitions that Peirce gives for a sign is as a medium for the communication of a form or habit embodied in the object to the interpretant, so as to constrain (in general) the interpretant as a sign or (in cognitive systems) the interpreter's behavior (De Tienne 2003; Hulswit 2001; Bergman 2000; Queiroz and El-Hani 2006b):

> … a sign may be defined as a Medium for the communication of a Form. […]. As a medium, the sign is essentially in a triadic relation, to its object which determines it, and to its interpretant which it determines. […]. That which is communicated from the object through the sign to the interpretant is a Form; that is to say, it is nothing like an existent, but is a power, is the fact that something would happen under certain conditions" (MS 793:1–3).

The notion of semiosis as form communicated from the object to the interpreter through the mediation of the sign allows us to conceive meaning in a telic, processual, non-substantive way, as a constraining factor of possible patterns of interpretative behavior through habit and change of habit.

Peirce's theory is compatible with the notion of self-organization, especially when we consider how semiotic habits stabilize communication within a system (a community) of utterers and interpreters, through self-correcting dynamics and circular relations within that community (Loula and Queiroz 2011). New habits of sign action emerge from previously existing sign action in a cultural evolutionary process. Out of semiosis that follows some stability and tendency of development, new habits become stabilized and begin to develop autonomously.

2.6 Active Semiotic Externalism

Peirce's thesis that mind is semiosis serves as a significant precursor to the ideas of the situated mind and distributed cognition (Atã and Queiroz 2014). We call this perspective active semiotic externalism. Unlike internalism, which confines mental contents to the intrinsic properties of an individual mind, externalism posits that mental contents, including intentional states, have a partial presence in the environment (Lau and Deutsch 2019). Active externalism goes further by suggesting that the environment participates in driving cognitive processes, and not only providing mental content (Clark and Chalmers 1998).[12]

Peirce's semiotic theory of mind entails a processual orientation to active externalism. In this framework, (1) sign action drives cognition, (2) this sign action extends beyond individual cognition, forming situated connections among individual minds and environments, and (3) this sign action is temporally distributed: it projects into the future tendencies derived from a past history. For active semiotic externalism the core unit of interest in cognitive science is a materially embodied, environmentally situated and temporally distributed process. As the mind is defined as a dynamic and dialogic process, it is not delimited by individual boundaries, whether selves or bodies. By the same token, meaning is not contained within a semiotic artifact, an individual's mental processes, the referent of the sign, or the medium of transmission. It emerges externally among organisms and resources in an ecosystem (Nielsen 2007); among species and environments in evolution (Kohn 2013); among individuals, groups, and institutions in society (Fusaroli 2011); among customs, traditions, and cultural artifacts in culture (Sinha 2009; among utterers and interpreters in communication (Ransdell 2003); and among parts of a

[12] It should be clear from the outset that what we call semiosis is not coextensive with high cognition or consciousness. In our view, semiosis is more ubiquitous and precedes the mind. Semiosis includes quasi-mind. Cognitive processes can emerge in a universe already imbued with external signs, as a complexification of triadic relations that were already present.

distributed cognitive system in cognition (Hutchins 2014; Zhang and Patel 2006; Kirsh 2009; Bjørndahl et al. 2014; Davies and Michaelian 2016).

Peirce's principle of continuity (or synechism; Esposito 2005) underscores the seamless, albeit mediated, interconnectedness of minds with other minds and their environments. Against any form of dualism, Peirce embraces a monistic view in which matter and mind are distinct yet intertwined states of the same process, as exemplified by the statement, "matter is effete mind" (CP 6.25). Here, there is no rigid ontological divide between mental and material processes. In the passage below, Peirce formulates his own version of a "parity principle" (Clark and Chalmers 1998; see Menary 2010), anticipating modern-day distributed cognition and extended mind hypothesis:

> A psychologist cuts out a lobe of my brain... and then, when I find I cannot express myself, he says, 'You see your faculty of language was localized in that lobe.' No doubt it was; and so, if he had filched my inkstand, I should not have been able to continue my discussion until I had got another, Yea, the very thoughts would not come to me. So my faculty of discussion is equally localized in my inkstand. It is localization in a sense in which a thing may be in two places at once. (CP 7.366)

Active semiotic externalism can be distilled into the proposition that all thought is in signs (MS[R] 654:3; EP 1:30; CP 1.538, 5.253, 5.265, 6.338). In it, the concept of mind aligns closely with that of a system capable of embodying semiosis: "a mind may, with advantage, be roughly defined as a sign-creator in connection with a reaction-machine" (MS [R] 318:18). In essence, mind is defined as a process of sign interpretation, a general pattern of triadic activity that extends across time and space, and cognition can be defined as a power to produce interpretants. It operates through the development of semiotic artifacts out of previous semiotic artifacts.

2.7 Trichotomies and Sign Classes

The phaneron is full of signs of many different types and which work in different ways—from physical entities to abstract general concepts, from signs which embody what they mean to signs which merely refer to it, and from signs which aim at multiple possible interpretations (such as an artwork) to signs which aim at very precise interpretations (such as a code of law). Peirce's system of categories offers an exhaustive typology of classes of signs to try to make sense of this variety of ways that signs act. So far in this chapter, we have used Peirce's triadic system of categories to define semiosis/cognition as a phaneroscopic thirdness which hierarchically involves a secondness (an object dyadically connected to a sign) and a Firstness (a sign as a monadic relation) as its logical components. Now we will explore how we can use the categories to further flesh out and analyze the nature of these component relations in order to generate a typology of sign classes.

Peirce's classification draws from distinctions of sign classes developed in ancient and medieval semiotics (such as type/token, term/proposition/argument, and relations of reason/real relations), but it systematizes and adjusts these early insights, grounding the classes of signs in the triadic categorical system. Peirce developed multiple classifications of signs (10, 28 and 66 classes) through various trichotomies (EP 2: 289–299 and 478–491). This has resulted in a significantly enhanced accuracy in understanding the relationships within semiosis (S–O–I). These trichotomies serve as aspects through which semiosis is observed and can be translated into essential questions (Houser 1991). Only a limited number of semioticians have explored Peirce's extensive typologies of signs, consisting of 10, 28 and 66 classes developed after 1903. These typologies remain structurally complex, and challenging to apply to real-world phenomena.[13] Furthermore, there exists a prevailing inclination to view these extended typologies as extravagant and unproductive conceptual tools, further complicating their acceptance. But in our opinion, these classifications should be regarded as a significant advancement in empirically modeling the morphological diversity of signs. In Peircean cognitive semiotics, the classes of signs should be regarded as maps of possible cognitive/semiotic states of a system and their interconnections, effectively providing a conceptual model of how cognition works.[14]

To arrive at the trichotomies that form the basis for the classification of signs, we can start by combining the two applications of the categories that we have already explored in the previous sections—(a) the elements of the phaneron may have the nature of feelings (monadic relations), reactions (dyadic relations) or laws (triadic relations); (b) a triadic semiotic relation involves a sign (the first of the relation), an object (the second) and an interpretant (the third). Combining these two properties, we can have feeling, reactions, or laws occupying the functional roles of sign, objects, and interpretants. In other words, the interpretant relation to the object and sign, the object relation to the sign, and the sign relation in itself, may each itself have the character of firstness, secondness and thirdness of the phaneron. This gives three fundamental trichotomies, which are the basis of the classification of signs in 10 classes: the sign relation may be a feeling, a reaction, or a law (first trichotomy); the object in relation to the sign may be a feeling, a reaction, or a law (second trichotomy), and the interpretant in relation to the object via the sign may be a feeling, a reaction, or a law (third trichotomy) (see Table 2.1).

The first trichotomy describes the morphological variety of the sign, or representamen, that is, the artifact that actively conveys meaning. It answers the question—What is the relation of the sign with itself? This relation may be genuinely triadic, and thus have the nature of a law, or it may be degenerate to the first degree,

[13] For an introduction to Peirce's extended classifications of signs, see: Farias and Queiroz 2000, 2003; Atkin 2013, 2016; Queiroz and Stjernfelt 2019.

[14] There is an author that should be mentioned, whose articles, especially in the 1990s, first shed light on the relationships between extended classifications of signs and phaneroscopy, Floyd Merrell (1991, 1994, 1996, 1997). According to Merrell, Phaneroscopy and Semiotics represent the same theory described in two different and complementary languages. Signs ('thought-signs' and 'event-signs' 'out there') in semiotics correspond to phanerons ('thought-signs,' 'in here') in Phaneroscopy. This equivalence, also suggested by Tursman (1987, p. 25), can be interpreted as follows—the results of semiotic investigations are, in many respects, the same as those obtained through phaneroscopy, albeit expressed in "complementary languages."

Table 2.1 Three trichotomies and three kinds of relations (CP 2.243)

	1st trichotomy	2nd trichotomy	3nd trichotomy
Monadic relation (S-dependent)	QUALIsign in itself, the sign is of the nature of an appearence (EP 2: 291)	ICON a sign which refers to the object merely by virtue of characters of its own (CP 2.247)	RHEME a sign which, for its interpretant, is a sign of possibility
Dyadic relation (S-O dependent)	SINSIGN in itself, the sign is of the nature of an individual object or fact (EP 2: 291)	INDEX a sign which refers to the object by virtue of some existential relation	DICENT a sign which, for its interpretant, is a sign of actual existence (EP 2: 292)
Triadic relation (S-O-I dependent)	LEGISIGN in itself, the sign is of the nature of a general type (CP 8.334)	SYMBOL a sign which refers to the object by virtue of some law	ARGUMENT a sign which, for its interpretant, is a sign of law (CP 2.252)

and thus have the nature of a reaction, a dyadic relation, or it may be degenerate to the second degree, and thus have the nature of a feeling, a monadic relation. When the semiotic artifact has the nature of a monadic relation, a feeling, it is called a qualisign, "any quality in so far as it is a sign" (CP 2.254). This accounts for when signification is conveyed by vague feelings, appearances and sensations. For example, the feeling of red is a qualisign. When the semiotic artifact has the nature of a dyadic relation, a reaction, it is called a sinsign, or a token, "an actual existent thing or event which is a sign" (CP 2.245). This accounts for all the instances when an individual object, fact, or event acts as a sign. For example, a particular red traffic light situated here and now is a sinsign. Because of the hierarchical nature of the categories, dyadic relations involve monadic relations, and thus a sinsign involves a qualisign—the particular red traffic light involves the pure feeling of red as one of its constituents that may be prescinded from many other feelings that accompany it, such as its shape, luminosity, texture, etc. Finally, when the semiotic artifact has the nature of triadic relation, a law, it is called a legisign, or a type, "a law that is a sign" (CP 2.246). This accounts for all the cases when a general pattern of concept acts as a sign. For example, the particular red traffic light situated here and now is just a situated instantiation of a general type—when I look at it, I don't see it for what it is here-and-now, but I see it in its general character as a pattern. Here again, because of the hierarchical nature of the categories, triadic relations are instantiated in dyadic relations, and thus a legisign has to be instantiated in several instances of sinsigns, thus forming a general pattern.

The second trichotomy is considered the "most fundamental classification of signs" (CP 2.275), and describes the morphological variety of the object in its relation to the sign. It answers the question: What is the relation between the sign and its object? If the relation is monadic (a triadic relation degenerate to the second degree), it has the nature of a feeling or quality. That is, sign and object are related by means of a quality they share. This is a relation of iconicity. Icons are signs which stand for their objects through similarity or resemblance, irrespective of any spatio-temporal physical correlation that S may have with an existent O. If the sign (S) represents the object (O) through a determinative relation based on analogy, signifying a shared quality between S and O, then S can be classified as an icon of O. The relationship between S and O is established by their shared identity in some aspects. Icons are very dependent on the material, form and structure of which they are made—"An Icon is a sign which refers to the object that it denotes merely by virtue of characters of its own, and which it possesses, just the same, whether any such object actually exists or not" (CP 2.247). For example, if someone points to the red traffic light and says "I bought a shirt that color", the relation between the red light and the shirt is one of iconicity. In this case, it doesn't matter, for the situated purpose of the communication, whether the shirt is real or fictional, the sign conjures up the red shirt as its object in that communicative situation.

In contrast, the semiotic artifact may be related to its object dyadically (a triadic relation degenerate to the first degree), a relation that has the nature of a reaction. This is the case when S is a sign of O by reason of "a direct physical connection" (CP 1.372) between them. In that case, S is really determined by O, in such a way that both must exist as events. This is a relation of indexicality, and S is said to be an index of O—"An Index is a sign which refers to the object that it denotes by virtue of being really affected by that object" (CP 2.248). The notion of spatio-temporal co-variation is the most characteristic property of indexical processes. The examples range from a pronoun demonstrative or relative, which "forces the attention to the particular object intended without describing it" (CP 1.369), to physical symptoms of diseases, photographs, weathercocks, thermometers. A red traffic light is indexically connected to the intersection whose traffic it helps to regulate. That is, when I look at a red traffic light, I know which intersection it refers to because it is physically situated right before it or next to it. A red traffic light doesn't only tell me to stop, it points to me where I should stop, and this pointing is an example of indexicality.

Finally, to complete the second trichotomy, we have the case in which the semiotic artifact is related to its object in a genuinely triadic way. This relation has the nature of a general law. Here, S is related to O only through I. The relation between sign and object is not due to similarity or analogy, nor to physical connection or spatiotemporal correlation. It is only realized in the process itself of interpretation, in the sense that this interpretation instantiates a habit. This is a relation of symbolicity, and in this case the semiotic artifact is said to be a symbol of its object. In a symbolic relation, the interpretant stands for 'the object through the sign' by a determinative relation of law, rule or convention (CP 2.276). For example, there is no analogy or physical correlation that explains the fact that the color red in a traffic light signifies "Stop" while the green signifies "Go"; these associations are based on habit, and thus are symbolic.

The second trichotomy may also be understood in terms of relative dependence of the S-O relation on S, O or I. That is, if the S-O relation can do without the third term (I), then it is a triadic relation degenerated in the first degree, and it is an index. If it can do without the second and third terms, then it is doubly degenerate, and it is an icon. Icons can do without the correlates of the triadic relation since they depend on their own qualities (CP 3.362). An index is a sign that would lose the character that makes it a sign if its object were removed, but it would retain this character if there were no interpretant (CP 2.304). If the relation depends on all the correlates, then it is a genuine relation and is a symbol.

The third trichotomy describes the morphological variety of interpretants in their triadic relation to sign and object. The relation between sign and object produces a new sign, and interpretant, which is also in relation to that object:

> A *sign*, or *Representamen*, is a First which stands in such a genuine triadic relation to a Second, called its *object*, as to be capable of determining a Third, called its *interpretant*, to assume the same triadic relation to its object in which it stands itself to the same object. (CP 2.274, 1902)

Such triadic relation *qua* triadic relation is described by the third trichotomy, while the second and the first trichotomies describe the S-O and S relations which hierarchically compose the triadic relation. The third trichotomy answers the question—What is the relation between the sign and its object for its interpretant? While the interpretant relation is triadic, it may either be genuinely triadic, and thus have the nature of a law, be degenerate to the first degree, and thus have the nature of a reaction, dyadic relation, or be degenerate to the second degree, and thus have the nature of a feeling, a monadic relation.

When the interpretant relation has the nature of a qualitative feeling, a monadic relation, it is a rheme, "a sign which, for its interpretant, is a sign of qualitative Possibility, that is, is understood as representing such and such a kind of possible object" (CP 2.250). This accounts for when the interpretative effect of a sign is merely a potential connection to an object. The sign doesn't tell the interpreter what to do or think, but merely presents a predicate. For example, someone who has never seen a traffic light before and doesn't know how it works might see the red light glowing and think to him/herself that maybe it is a sign of danger. When the interpretant relation has the nature of a reaction, a dyadic relation, it is a dicent sign, or dicisign, or proposition: "Dicent sign is a sign, which, for its interpretant, is a sign of actual existence" (CP 2.251). This accounts for when a sign dyadically compels a certain interpretative response, telling the interpreter what to do or think. For example, when the light turns red, it effectively tells drivers "Stop here and now!" When drivers stop, they are acting on the basis of a dicent sign. Finally, the interpretant relation may also be genuinely triadic, and thus have the nature of a law or habit. In this case, it is an argument, a "sign which, for its interpretant, is a sign of law" (CP 2.252). This accounts for cases when a sign doesn't only tell an interpreter what to do or think, but it justifies it on the basis of an inference. Abductive (or hypothetical), deductive and inductive inferences are three kinds of arguments. For example, the person who doesn't know how traffic lights work asks a driver why he/she has stopped, and they respond, explaining what a red traffic light means, what one is supposed to do when the light turns red, and pointing out the red glowing light in front of them to show that this is just the case now. Such a response has the form of a deductive argument (demonstrating that a case pertains to a general rule).

It is a genuine triadic relation because it is wholly dependent on the interpretant to bind together sign and object by means of a law or habit.

We have, then, three trichotomies to express the morphological variety of semioses—the sign relation may be a qualisign, a sinsign or a legisign; the object relation may be an icon, an index, or a symbol; the interpretant relation may be a rheme, a dicent sign, or an argument. If we combine them, we form sign classes; for example, a legisign, which is also a symbol, is also an argument. Or yet, a legisign, which is also a symbol, which is also a dicent sign. We can represent these combinations by a series of numbers, each referring to a category of a trichotomy. Thus, the sign class 321 is a legisign (first trichotomy is a thirdness: 3__), an index (second trichotomy is a secondness: _2_), and a rheme (third trichotomy is a firstness: __1). Such a class would correspond to when a sign is a general law or pattern, which signifies by means of a physical connection, but which merely presents a quality or feeling for interpretation, such as a demonstrative pronoun, considered alone (the sign is a type word, which signifies by means of a spatiotemporal correlation with a noun which appears elsewhere in a body of text, but which in itself is not a proposition, but merely conveys a possible object).

How many classes are there? If we combine the three possibilities for the three trichotomies, that would give us 27 classes of signs (3 times 3 times 3). However, there are hierarchical constraints that preclude most of these classes. The trichotomous divisions are structured according to an order of assumption, associated with the categories: character of sign presentation (firstness), character of sign representation (secondness), and interpretive power of the sign (thirdness). According to Liszka:

> the presentative aspect of a sign can only be combined with representative aspects which are equal to or lower than the presentative's phenomenological type; the representative aspect of the sign can only be combined with interpretative aspects which are equal to or lower than the representative's phenomenological type. (Liszka 1996, p. 45)

Savan (1987–1988, p. 14) uses the term "qualification rule" (which describe the fundamental concept known as praecisio, CP 1.353, 1.549, 2.428, 5.449), which governs the way responses from trichotomies are combined, stipulating that each relation type "qualifies" the subsequent one. Simply put, a "first" can only be qualified by another "first"; a "second" by both a "first" and a "second"; and a "third" by a "first," a "second," and a "third" (Savan 1987–1988, p. 14). Savan further clarifies that within any sequence, a category member can be followed by one equal or smaller—a "first" by another "first"; a "second" by a "first" or a "second"; and a "third" by a "first," a "second," or a "third." For example I see the red traffic light as a symbol for stop, I am not anymore considering the traffic light as a sinsign (this physical artifact, here and now) only, but already as instantiation of a general type (this physical artifact is a "traffic light", and I know how it works in general, and thus I can tell it is telling me to stop at this intersection). Thus we cannot have a sinsign which is a symbol, being a symbol already implies a legisign.

This constraint precludes any class in which a certain number is followed by a bigger one (e.g., 123, 112, 233, 213, etc.). This reduces the number of available classes, from 27 to ten: 111, 211, 221, 222, 311, 321, 322, 331, 332, 333. Weiss and Burks (1945, p. 387) contend that this principle extends beyond the ten classes, broadly determining the number of classes achievable from a set of trichotomies. Their proposed formula, (n + 1) (n + 2) / 2 (where n represents the trichotomy count), illustrates this concept: three trichotomies yield ten sign classes, six generate 28, and ten result in 66 sign classes (as stated by Peirce in a letter to Lady Welby, SS 85; EP 2: 481). By following this formula, considering only one trichotomy leads to a division into three sign classes, as observed in "On a New List" (W 2: 49–59), and other pre-"Syllabus" texts such as "One, Two, Three: Fundamental Categories of Thought and of Nature" (W 5: 243) and "An elementary account of the logic of relatives" (MS 585; W2: 379), where Peirce exclusively explores the division into icons, indices, and symbols.

The cross-relations that satisfy the constraints, or the 10 classes of signs, are (EP 2:294–296; CP 2.254–263) (Fig. 2.1): (1) a Qualisign is a quality "in so far as it is a sign"; its object is interpreted as being of the same nature; (2) an Iconic Sinsign is a sign-event interpreted as possibly standing for its iconic object; (3) a Rhematic Indexical Sinsign, is a sign- event interpreted as possibly standing for an event; (4) a Dicent Sinsign is an sign- event interpreted as spatio-temporally standing for another event (index); (5) an Iconic Legisign is a type, or a law, a "regularity of the

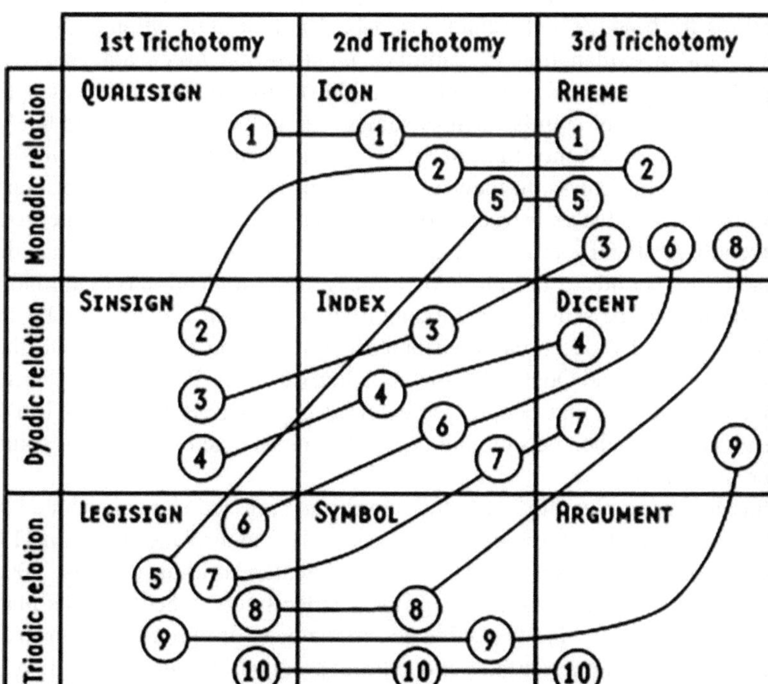

Fig. 2.1 The 10 classes of signs as a system of cross-relational classes. The paths correspond to the possible compounds of relations. (Figure based on Merrell 1996, p. 8)

indefinite future", interpreted as possibly standing for its object (icon); (6) a Rhematic Indexical Legisign is a type interpreted as possibly standing for its object (another event); (7) a Dicent Indexical Legisign is a type interpreted as spatio-temporally reacting with its object (another event); (8) a Rhematic Symbol is a type interpreted as possibly standing for its object (law); (9) a Dicent Symbol is a type interpreted as physically standing for its object (law); (10) an Argument is a type interpreted as semiotically standing for its object (law).

The importance of this classification must be emphasized. The morphological space of semiotic processes in which cognitive systems are embedded include proto-symbols and variations of indexical signs, beside symbolic and iconic processes. And there is no way to describe these processes with some accuracy examining only the sign-object relationship. Peircean mature typologies provide a detailed description of several inter-related aspects involved in semiosis including the intrinsic nature of signs and the effect on the cognitive systems.

2.8 Semiosis Is Temporally Distributed

Peirce's semiotics simultaneously offers a processualist theory of meaning and a processualist theory of mind. This philosophical position has three fundamental implications for how we understand semiosis. The first implication is that the fundamental unit of explanation in semiotics is "sign-in-action", not "sign" (Fisch 1986, p. 330). Quoting Max Fisch (1986, p. 330): "the fundamental conception of semeiotic is not that of sign but that of semeiosis; and semeiotic should be defined in terms of semeiosis rather than of sign…". The focus here is on semiosis—the action of signs—in contrast to theories of meaning which focus either primarily on the sign itself (formalist and structuralist approaches) or on sign-users (neurocognitive, psychological and sociological approaches). The second implication is that the mind itself corresponds to a semiotic process. A third implication is that mind/semiosis is external (in the sense of active cognitive externalism). Peirce can be considered a precursor of modern-day distributed cognition and extended mind theories (Paolucci 2011; Atã and Queiroz 2014). As Kirsh (2009, p. 297) recalls, "Peirce first mentioned this idea – that people use external objects to think with – in the late nineteenth century, when he said that chemist think as much with their test tubes as with pen and paper."

In classical examples of distributed cognition, sign action is distributed within spatial contexts: boats (Hutchins 1995a), cockpit of an airplane (Hutchins 1995b), research laboratory (Nersessian et al. 2003), theater building (Tribble 2005). In these examples, the distribution of cognition is decisively conceived as a matter of spatial locus: cognition is not only located in the head of individuals, but in spatial environments of cognitive artifacts and cultural practices. In our approach, the focus of description of distributed sign action is not spatial, but temporal. It is not only the case that a sign cannot simply be described as spatially contained within one's skull, it is also (and more centrally) the case that a sign cannot be described as temporally contained within one event or episode. The most fundamental explanatory unit in active semiotic externalism is sign action, a temporally distributed process. Semiosis self-organizes in time, in a process that continuously entails the production of more signs. Even when the focus of observation is effectively instantiated sign action in a single event, the sign acts as a temporally distributed process in which a regular tendency towards certain future outcomes emerges from a history of sign interactions.

Semiosis is a dynamic process, with signs continuously translating into other signs, in time. The mode of existence of a sign is to produce interpretative effects which are new signs. A single person thinking to herself already embodies a dynamic process of semiosis. This temporal distribution of semiosis is a central aspect of Peirce's semiotics:

> To say, therefore, that thought cannot happen in an instant, but requires a time, is but another way of saying that every thought must be interpreted in another, or that all thought is in signs. (CP 5.253)

Semiosis is temporally distributed, a time-dependent action of signs which develops in several historical and evolutionary scales. To describe a habit of sign action is to describe a dynamic process involving past and future: a relation, developing in time, between a cumulative generalization of a past history of sign actions and a likelihood that this history will somehow regulate future sign actions. In this framework, stability of semiosis in time is not understood as an instantiation of representational states in a system but as regularity of sign action throughout communication. A habit doesn't entail stasis, but regularity throughout change (Fig. 2.2).

sign (S) as medium for the communication of a form embedded on the object (O) to the interpretant (I). A sign can be an external artifact, as well as a mental representation. The object is that for which the sign stands. The interpretant are the effects caused by the relation between sign and object to an interpreting mind.

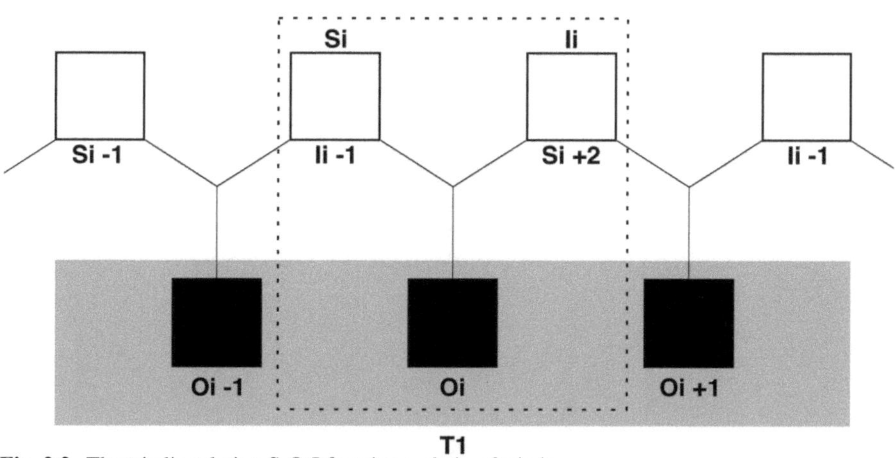

Fig. 2.2 The triadic relation S-O-I forming a chain of triads

Chapter 3
Semiosis as an Emergent Process

3.1 Signs of Emergence

The concept of emergence refers to a non-linear relationship between levels (organizational or observational) and/or ontological domains of a process, where the properties and rules of the emergent level or domain are qualitatively different from those of the constituent parts or previous phase of the process. This suggests that reality operates across different levels and/or domains with different local properties and rules. For instance, biological processes emerge from chemical and physical interactions, societies emerge from individual interactions, and mental activity emerges from neurobiological activity. Emergence is associated with irreducibility and unpredictability, meaning that the properties of a system cannot be deduced from those of its parts, or alternatively, the emergent behavior cannot be explained in terms of the behavior of the parts. As an explanatory framework, emergence attempts to account for true novelty within a naturalistic view of the universe.

The potential of emergence in the field of semiotics lies in its ability to provide an understanding of how sign processes are composed of nothing else than material interactions between organisms, artifacts and environments, and yet they possess unique properties which cannot be found in the material interactions that comprise them.[1] Semiotic emergence requires, thus, both a stance of scientifically sound naturalism, and of respect to the creative capacity of semiosis to self-regulate and constitute niches, conceptual spaces, and/or ontological domains which have to be understood in terms of their own unique rules and properties.

Using the terminology of Sartenaer (2016), the promise of emergence in semiotics is to clarify how sign processes can be both continuous and discontinuous to the

[1] To see a publication on the relationship between theories of emergence and pragmatism, see: Baggio and Parravicini (2019).

P. Atã, J. Queiroz, *Mind as an External Sign: A Semiotic Externalist Philosophy of Cognitive Science*, Synthese Library 531,
https://doi.org/10.1007/978-3-032-08988-5_3

material interactions that compose them. In our description, sign processes are embodied-situated in the material and cultural environments of organisms and artifacts, but they are also temporally distributed. Temporal distribution, here, refers to how sign processes inextricably operate across different temporal scales of interpretant production, comprising generalized pasts, anticipated futures, and multiple possible ways of connecting them together.[2]

The concept of emergentism is better suited to a processualist ontology than a substantialist framework (Bickhard 2003; Campbell and Bickhard 2011). Emergence can be challenging to explain within a substance-oriented view, as it requires accounting for how physical matter gains causal powers from non-physical (organizational) processes (see Kim 1999), such as the problem of explaining mental activity solely from biophysical properties. In contrast, a process-oriented understanding of emergentism focuses on how causal powers are held by organized processes through and through. A stable substance is a result of organization, meaning that different organizations can generate different causal powers, without any violation of naturalism and without positing new kinds of substances. A process-oriented view of emergence prioritizes interlevel and interdomain relations and organizations over substances and units. As a result, process-oriented emergentism is concerned with explaining how sign processes can develop and grow *across* interactions of organisms, artifacts, and environments, rather than on how physical matter can possess some sort of pre-formed semiotic potential. We consider Peirce's description of phenomena as having the nature of relations, and of cognition as a triadic kind of relation, to be a strong candidate in the programme of process-oriented emergentism.

Emergence can be conceived in a variety of ways (Stephan 1998, 1999; Kim 1998, 1999; Sartenaer 2016). To begin with, emergence can be conceived as a diachronic or synchronic phenomenon. Diachronic emergence proposes to resolve the continuity-discontinuity relation between bases and emergents with the help of time. A base domain goes through a process of emergence of new properties and/or substances, thus becoming something else. Bases and emergents do not coincide in time. Here, emergence is a process of transformation into something new. Bases and emergents are discontinuous when we consider them as initial and end states of a process of transformation, but yet continuous when we consider that these states are merely the terminals of the same process (temporal development). Thus diachronic emergence fits in well with a flat/transformational conception. One doesn't need to pose levels of observation or organization, but merely distinct domains which are separated in time: a base domain which transforms in an emergent domain. In contrast, synchronic emergence stresses the coincidence, in time, of continuity and discontinuity between bases and emergents. At a given moment, a process exhibits both base components and emergent properties. They are continuous in the sense that they necessarily share the same temporal slice (without the base components,

[2] The idea that semiosis operates on several temporal scales appears in one of the articles that Peirce published between 1867 and 1868, relating time, sign, and mind. We will come back to this. For a detailed approach, see Murphey (1993).

there wouldn't occur the emergent properties), and yet they are discontinuous in the sense that there is a qualitative difference between them. In this case, one has to pose a hierarchical distinction between levels of a process. A part of a process possesses a different set of properties when considered at the base level or at the emergent level. This brings us to the third distinction, between weak or strong emergence. These levels may be conceived as levels of observation or levels of organization, depending on whether the emergence is conceived more as an explanatory or epistemological (weak) phenomenon or as an ontological (strong) phenomenon. The same distinction can be made also for diachronic and transformational emergence, in the sense that base and emergent domains can be conceived as either epistemological domains or ontological domains.

Conforming to two main varieties of emergence theories, semiotic emergence can be conceived in a more hierarchical, synchronic and holistic sense, or in a more flat, diachronic and transformational sense. In a hierarchical sense, semiosis is a triadic, S-O-I relation which synchronically involves dyadic and monadic relations as components. There is a hierarchy of presupposition and containment of relations in the S-O-I model: signs relations, which are Firsts of the triad, are contained and presupposed by sign-object relations, which are Seconds of the triad, and these by sign-object-interpretant relations, which are Thirds. The sign-object-interpretant relation is thus not a flat relation involving three elements, but a nested organization that can be represented thus (S-O-I(S-O(S))). Thus, in everyday life, by means of local reactions and direct feelings, we deal with general habits, beliefs and concepts, which emerge as holistic organizations of reactions and feelings, and which are present in experience and cognition together with those reactions and feelings but in a different hierarchical level. However, at the same time, semiosis has to be understood as a non-linear dynamic process, which allows semiotic emergence to be conceived in flat/transformational terms. Semiosis is a process that successively produces effects. This production does not develop in a straightforward, predictable manner: conversations take unexpected turns, differing reception defies communicative intentions, scientific and artistic revolutions suddenly change expectations, etc. In most cases, semiosis is non-linear. Interpretants emerge dynamically and are influenced by a complex array of conditions, rather than being determined linearly by the sign and its object. As a result, the process of semiosis is continually producing interpretants and re-evaluating interpretant production. This non-linear flow of interpretant production highlights the importance of situated conditions in shaping the meaning of a sign. But it also underscores the idea that meaning is not fixed nor predetermined by these conditions, and instead depends on a complex interweaving of past interactions (which are generalized in the form of a habit) and future interactions (which are anticipated).

3.2 Emergence and Continuity

Emergence can be understood as a continuity-discontinuity process. Sartenaer (2016) argues that theories of emergence aim at reconciling two seemingly contradictory claims: the continuity or homogeneity between emergent entities and their bases, and the discontinuity or heterogeneity between them. Emergent processes are seen as being constituted by their bases, while simultaneously being distinct from them. Sartenaer points out that theories of emergence attempt to achieve this reconciliation by according different scopes to continuity and discontinuity: an emergent process can be continuous to its base in some aspects, but discontinuous in others, such as exhibiting compositional continuity while also displaying explanatory discontinuity.

The characterization of emergence involving the unity of continuity and discontinuity allows us to understand emergentism, negatively, as an epistemological position that rejects both monistic reductionism, the view according to which all reality is reduced to the same underlying principles, and pluralism, the view according to which reality is divided or separated in different parts that respond to different principles, the most famous variety of which is dualism. It is in this context that emergentists look for a way of conceiving reality as both continuous (and, thus, in some sense monistic) as well as discontinuous (and, thus, in some sense, pluralistic). Thus, for semiosis to be emergent, its emergents and bases (however we understand them) need to be both continuous and discontinuous to each other.

Peirce's philosophy entails a form of non-reductive monism. In epistemology, for example, this is expressed in the form of pragmatism. Peirce's monism is underpinned by a doctrine of continuity, called synechism (Esposito 2005). But synergistic continuity doesn't entail reductionism or determinism. Synechism conceives of processes that are continuous and open to chance and unpredictability. In fact, Peirce's characterization of synechism describes a form of continuity which can only be continuous through openness to real novelty. It is not, as reductionism would have it, a pre-formed continuity, already realized in the past, and from which variety is derived deductively. Rather, it is a pragmatist continuity *in futuro*, not preformed, and whose development is not only deductive, but also abductive and inductive: an emergent form of continuity.

Peirce's principle of continuity, or *synechism*, is primarily a methodological principle, "a maxim to look for connections and continuous strata between seemingly disconnected entities or events" (Esposito 2005, par. 1). Although Peirce construes synechism as a methodological principle, it points to a metaphysical hypothesis: "On the metaphysical side synechism is a hypothetical description of a tightly woven universe, a universe woven not within layers of the same kind of reality but between layers in a scalar fashion" (Esposito 2005, par. 1). Synechism is indissociable from thinking in terms of triadic relations. The synechist is skeptical of using dyadic principles as the categorical structure for understanding phenomena. Dyadic relations are relations of struggle between sharply separated parts.

Synechism is a skepticism towards any sharp borders or cuts that separate very neatly between dissociated modes of being:

> There is a famous saying of Parmenides, (...) "being is, and not-being is nothing." This sounds plausible; yet synechism flatly denies it, declaring that being is a matter of more or less, so as to merge insensibly into nothing. (EP 2: 2)

Against the dualism of dyadic distinctions, continuity is a property of Peirce's triadic system of categories, especially the category of Thirdness. It takes the form of dynamic generality: a continuous phenomenon is a process that develops in time, its continuous temporality is itself its continuity.

In Peirce's philosophy, the notion of continuity is derived from the study of mathematical continua (Buckley 2012). As is often the case in Peirce's philosophy, this formal investigation provides the fundamental tools for epistemology (where the principle of continuity forms the regulatory basis for pragmatism, and thus by extension to semiotics) and finally for metaphysics (synechism as a metaphysical hypothesis). The most immediate contexts of application of this mathematical-logical concept, then, is to be found in epistemology and semiotics, and then metaphysics. Stjernfelt (2007, p. 6) lists six points whose metaphysics necessitate continuity: "(1) the intensional meaning of general concepts (the inexhaustibility of continuous extension); (2) the embeddedness of actually existing objects and occurring events within a horizon continuum of potentiality; (3) realism as to general tendencies (as opposed to the powder of unconnected singular events without continuity); (4) continuity of research – from its infinitesimal beginnings long before science and to its converging end point in truth; (5) fallibilism as implied by the vagueness inherent in continuity; and finally: (6) diagram manipulation as basically continuous and hence able to mirror real continuity."

Peirce's work on mathematical continuity aims at proving that a continuum cannot be formalized as a line built up from points (Stjernfelt 2007, p. 4). For Peirce, a geometrical line cannot be understood as an arithmetic line. Or, in other words, one cannot reduce a continuous geometrical dimension to any countable series of points. We are dealing here with a comparison between infinities: even if one defines an infinity of individual points, there is still a greater infinity in the continuous geometrical space between points: "A true continuum is something whose possibilities of determination no multitude of individuals can exhaust" (CP 6.170). Peirce disagreed with Cantor's definition of continuity, which implied conceiving geometrical continuity as a set of discrete points. The basis for both Cantor's and Dedekind's arguments involves the arithmetization of continuity. According to Kent (1997, p. 449), "Peirce disagreed with Cantor's definition of continuity, which implied that geometrical continuity has to be thought of as a set of discrete points. Peirce claimed that Cantor's definition could not be used in topology and~topology (as it was then conceived) was thought to be pre supposed by all of geometry (CP 4.219). He then proposed an alternative definition of a continuum which contains no discrete points. Such a definition is consistent with the theory of real infinitesimals which Peirce wanted to retain."

This principle translates to a thesis about the relation between firstness, secondness and thirdness. Against the gravitational pull of dualism and nominalism, both of which are associated with secondness, the category of actuality, particulars and dual oppositions, synechism emphasizes the irreducibility of thirdness. It states that no infinity of actual particulars can exhaust general habits. A particular experience that actually takes place is merely part of a temporal growth that connects together uncountable other actual experiences so as to form an inexhaustible variety of patterns of occurrences. Thus, continuity is the basis of fallibilism and pragmatism, as it underlies the idea that no collection of data-points will be able to exhaust the predicates or the potential effects expressed by an object of investigation:

> The principle of continuity is the idea of fallibilism objectified. For fallibilism is the doctrine that our knowledge is never absolute but always swims, as it were, in a continuum of uncertainty and of indeterminacy. Now the doctrine of continuity is that all things so swim in continua. (Untitled manus., c. 1897, CP 1.171)

While dyadic principles concentrate on the joints that separate and define processes, synechistic continuity suggests that processes will resist absolute measurements and definitions, so that carving out reality at its joints is merely an act of approximation. Synechism assumes that even in the face of very precise measurements, there is always an inherent uncertainty, an impossibility of nailing down phenomena with perfect accuracy, and, likewise, that phenomena cannot be "frozen" into timeless propositions. It asks us to understand being in terms of degrees of regularity, or approximative tendencies towards regularity, instead of perfect regularities, and thus, by the same token, it conceives of knowledge as dynamic.

In his later years, Peirce reconceived the notion of continuity in terms of temporality. In fact, he seemed to draw an equivalence between continuity and time. That is, to say that anything is continuous means nothing else than to say that time is unbroken. We should understand continuity as "the relation of the parts of an unbroken space or time" (CP 6.168). He admits a circularity in this definition:

> In endeavoring to explicate "immediate connection," I seem driven to introduce the idea of time. Now if my definition of continuity involves the notion of immediate connection, and my definition of immediate connection involves the notion of time; and the notion of time involves that of continuity, I am falling into a circulus *in definiendo*. But on analyzing carefully the idea of time, I find that to say it is continuous is just like saying that the atomic weight of oxygen is 16, meaning that that shall be the standard for all other atomic weights. The one asserts no more of time than the other asserts concerning the atomic weight of oxygen; that is, nothing at all. (CP 4.624)

Time is recast as the standard for the notion of continuity. However, conversely, the idea of time becomes nothing else than the idea of continuity. Continuity is not a property of time, nor temporality is a property of continuity, but rather time and continuity define each other circularly. In this, time ceases to be a grammatical subject and becomes merely a predicate. That is, it becomes meaningless to predicate unbrokenness to time, rather, temporality is itself the predicate of unbrokenness. Time is the mode of connection which defines continuity.

Peirce offers a scenario of application of the notion of continuity in the phenomenology of time passage. A number of approaches, in recent decades, have pictured cognition as dynamical, occurring in real time and dealing with a continuous and probabilistic stream of percepts—in contrast to models that assume discrete and well-defined mental states sandwiched between inputs and outputs (for example, Port and Van Gelder 1995; Ward 2002; Chemero 2001; Dove 2016). Also Peirce proposed a theory of mind as dynamic activity, applying his principle of continuity to the issue of mental representations: "At no one instant in my state of mind is there cognition or representation, but in the relation of my state of mind at different instants there is" (CP 5.289). Peirce's dynamical conception of mental representations suggests that semiosis is indistinguishable from a kind of temporal development.

Peirce argues that ideas cannot be sharply in the present or in the immediate past. The present needs to be constantly overlapping with the immediate future and past: "feeling which has not yet emerged into immediate consciousness is already affectible and already affected" (W 8: 150). A logical description of this process is given by Peirce in the form of continuous inferential acts that interweave past, present, and future in terms of general objects: "[W]hen a feeling emerges into immediate consciousness, it always appears as a modification of a more or less general objects already in the mind." In Peirce's account, it is a natural tendency of semiosis (a "law of mind") to weld together, through inferential acts, vague immediate experiences into more general ideas: "A finite interval of time generally contains an innumerable series of feelings; and when these become welded together in association, the result is a general idea" (W 8: 149).

The operation of a mind produces phaneroscopic time, underpinned by a temporal direction of development of an inferential flow, in which a thought or experience in the future is suggested by a thought or experience in the past. This is a diachronic process, then. However, this production of a temporality is also, at all times, a comparative co-determination between general ideas, actual occurrences and vague feelings, and thus it is also a synchronic process.

In this context, semiosis can be described as dynamic operation: it triadically connects a present feeling with several occurrences in the past and with an anticipated future through generalized patterns, or habits. (Generality, in this context, should not be regarded as an attribute of an abstract content within an individual's mind but rather as a fundamental mediating capacity.) To say that semiosis entails dynamic experience and thought is equivalent to saying that semiosis is continuous. The phaneron is a continuous whole: not a disjointed collection of feelings, reactions and habits, but a dynamic trajectory in which events unfold according to regular patterns. The holism of the phaneron can be experienced as the continuous experience of time passing. In Synechism, this temporal holism is indissociable from the holism of general concepts. It is because we can generalize feelings and events and experience them as comparative transformations of general ideas, that we can experience dynamical time, and not just disjointed feelings, occurrences and concepts.

On the other hand, for semiosis to be emergent, it is also necessary that this continuity is not absolute, that the regular patterns that mediate the dynamic unfolding of events are open, not set in stone. Synechism does not imply the absence of transformation, nor deductive preformationism. As exemplified by the notion of fallibilism, synechistic continuity is a kind of continuity which can only exist because it lacks absolute determination. It is continuity that involves and requires chance and openness to novelty. In this sense, semiotic emergence does not exactly confer distinct scopes to continuity and discontinuity, as in the analysis of Sartenaer. Rather, it rejects any dualism between continuity and discontinuity, replacing their dyadic opposition for a triadic mediation.

3.3 What Emerges in Semiotic Emergence?

Semiotic emergence describes the emergence of dynamic and holistic Thirdnesses out of Secondnesses and Firstnesses. However, as we have seen in the last chapter, there are at least three ways in which the triadic system of categories is applied so as to produce a model of cognition: to differentiate and define semiosis as a triadic relation, in contrast to dyadic and monadic relations, to differentiate and define the three relative positions within a triadic relation (First, Second, and Third, or sign, object, and interpretant), and to differentiate and define the component-relations that occupy each of the these three relative positions (the sign, the object and the interpretant may each be a Quality, Reaction or Law). If we associate the notion of emergence to Thirdness, then each of these applications of the notion of Thirdness to cognitive semiotics supports a different specific understanding of semiotic emergence: emergence of semiosis (triadic relations) from non-semiotic relations (reactions and feelings); emergence of interpretants (thirds in a triadic relation) from signs in connection to objects; and emergence of triadic sign classes (legisigns, symbols, and arguments) from dyadic and monadic sign classes. We explore each of these understandings below.

The first possible answer to the question of what are the bases and emergents in semiosis is that semiosis itself emerges from the reactive experience of external facts (dyadic experience, or secondness), and these in turn from qualitative experiences (monadic experiences, or firstness). Peirce's phaneroscopy, or analysis of experience, consists in, starting from mental experience (which is the basis for the category of thirdness), extracting or separating from it the reactive experience of external objects (which is the basis for the category of secondness), and of qualitative feelings (which is the basis for the category of firstness). In this analysis, dyadic relations can be prescinded from triadic relations, and monadic relations can be prescinded from dyadic relations.

Semiotic emergence can be formulated as a process that moves in the opposite direction as the phaneroscopic analysis. That is, while the analysis extracts secondness as a component of thirdness, and firstness as a component of secondness, emergence would be the process in which thirdness emerges from secondness, and secondness emerges from firstness, which each of these steps corresponding to a change in the organization of qualitative experience. That is, secondness is a reorganization of the qualitative feelings found in firstness, which transforms them from feelings into reactions, and thirdness is once again a reorganization of those reactions and feelings, which transforms them into semiosis.

The first step of this process is the emergence of experience of reactive, existent, objects (secondness) as a certain organization of experiences of qualitative feelings (firstness). The key for this transition is that some qualitative feelings cannot be commanded by will, cannot be imagined away, or yet dreamed into experience, and thus are not mere feelings in one's mind, but feelings partly commanded by something else. This has to do with how qualitative feelings follow one from another and/or interrupt one another. The decisive relation in this form of organization is of *involvement*: the same existent object involves several qualitative feelings, and commands them in a way that they manifest with a certain compulsory organization which cannot be imagined away. Thus, out of several qualitative feelings, and by virtue of how they are organized, there emerges the experience of reactive, and thus existent, objects.

The second step of this process is the emergence of the experience of semiosis (mediation, thought, cognition, generalization) as a certain organization of the experience of existent objects. The decisive relation in this form of organization is of *instantiation*. An experience of an existent object is taken to be an instantiation of a more general regularity. Thus, the direct experience in front of my eyes becomes mediated by other, past, experiences, which we subsume and generalized in a class ("chair"). At this emergent level, we cannot experience an object only as an existent that reacts to my actions, but as an instance of a general class. Whether the object mediates a class which we already recognize, or a class which we don't know yet—in which case we experience the object as something curious or exotic to our knowledge—, that doesn't change the fact that we experience the existent object in front of us through the mediation of a general concept. The mode of organization here is one in which dyadic experiences are bound together by regular patterns, so that a certain dyadic experience is taken as an instantiation of such a pattern.

In this description, semiosis emerges in a phaneron that is ultimately composed of qualitative feelings—the monadic relations of the phaneron. First, by virtue of how several feelings are organized together, there emerges the experience of an existent fact that involves those feelings. Second, by virtue of how existent facts are organized together, there emerges the experience of general classes, signs or thoughts which are instantiated by those existent facts.

Another possible framing for the notion of semiotic emergence concerns the emergence of interpretants in a sign triad. Emergence, here, is the process through interpretative effects emerge in cognition and communication. We experience the emergent nature of interpretants as a pervasive condition of ordinary life. For instance, as communicators, we have to assume that we cannot predict exactly how an interlocutor will react to what we say, or inversely, that we might not react the way others would expect. Communication defies a linear trajectory and is often subject to complex translations. Interlocutors bring their own interpretative tendencies and rules into the conversation, which may not align with those held by others. Interpretative tendencies and rules that condition a cognitive system might not exist as such in another. The consequence is an inherent openness in communication that we try to account for and deal with using metasemiotic strategies for maximizing trust and common ground, for keeping communication flowing even in face of imperfect understanding, and for translating across different rules and contexts of interpretation.

In Peirce's logic, positions of a relation are ordered, and the order matters. The hierarchical ordering of the semiotic triad can be used to model the process through which triad itself emerges. This ordering implies a hierarchy of presuppositions, and also a development. The First of the relation, the sign, can be considered in itself, and regardless of the relation. The Second of the relation, the object, can be considered as an external factor that constrains the behavior of the First. The Third of relation, however, which is the interpretant, is the term that most fully depends on the relation. It is produced only in the context of the relation and is triadically irreducible. It emerges in the semiotic process that it integrates, and so it is incomplete at the start of this process.

The emergence of the interpretant as a Third of the triadic relation is not sudden, but includes degrees through which the conditions are set for the interpretant (and with it the process itself), to acquire genuine triadic status. At First, the interpretant is a potential already represented by the sign, the immediate interpretant. The immediate interpretant cannot be considered emergent, because it is already preformed in the sign itself. Secondly, as the sign itself comes into real connection with a dynamic object, the potential interpretation has to transform into an actual effect: the immediate interpretant is transformed into a dynamical interpretant. However, this production of an actual interpretative effect is not a direct determination of the dynamical object over the sign (in which case it would be merely a dyadic product). It is mediated by a regulatory factor: a tendency of development, or habit, that anticipates to the present the logical conclusion of the semiotic process in the future. This is what we described previously as the temporal distribution of semiosis: a logical situation in which a process in development only gets to be fully realized by an anticipation of itself in the future. This self-organizing principle is the Final interpretant.

Lastly, semiotic emergence can be thought of in the sense that sign classes which are more habitual emerge from sign classes which are less. Here, general, regular and stable meanings emerge from particular, irregular and unstable sign action: beliefs emerge from hypotheses and evidences, widespread cultural norms emerge from local occurrences, codified vocabularies of signs emerge from varied everyday usage, and so on. In terms of the 10 classes of signs, this means legisigns emerge from sinsigns and qualisigns, symbols emerge from indices and icons, and arguments emerge from dicents and rhemas, or, in overall, arguments (333) ultimately emerge from qualisigns (111). Emergence, here, can also be conceived as a process that runs opposite to the hierarchical relations of involvement and instantiation, since legisigns are instantiated in sinsigns, symbols are instantiated in indexes, arguments are instantiated in dicisigns (dicent signs), and sinsigns involve qualisigns, indices involve icons, and dicents involve rhemas.

The emergence of Thirdness in sign classes can be described as inquiry. Inquiry is a process that embodies a tendency towards the regularization of interpretant production. Through inquiry, a mental activity seeks to self-organize in the form of beliefs, that is, rules of action (W 3:263), or habitual connections of ideas (W 3:107). It typically involves three inferential steps: abduction, deduction and induction. This is a process that may recruit different classes of signs. For example, abduction is associated to perceptual clues in an environment, and thus with iconicity (Paavola 2005, 2011), deduction depends on diagrams (iconic legisigns) for deriving consequences from premises, induction depends on instantiated occurrences (sinsigns) that provide evidence (such as indices) for a general rule (Stjernfelt 2023).[3]

3.4 Synchronic or Diachronic Emergence

Diachronic and synchronic emergence (see Stephan 1998, 1999) refer to different phenomena, or yet to different aspects of emergence. Both approaches recognize the continuity and discontinuity of emergent properties to their bases, but they differ in their temporal scope. Diachronic emergence is interested in the appearance of novel properties over time. Emergents are continuous to their bases in the sense that they develop out of them, but they are also discontinuous in the sense that this development is not predetermined, but rather an unpredictable outcome of the complex interactions between the components of the system. In synchronic emergence, on the other hand, bases and emergents occur simultaneously and exist alongside one another. This means that at any given time, emergents are synchronically continuous and discontinuous to their bases; that is, they are determined by their bases, yet they are still considered to be irreducible to them. The concept of irreducibility suggests that emergents cannot be fully explained or broken down solely in terms of their bases. Instead, emergents are seen as distinct and unique phenomena that arise from the interaction of their underlying components. Synchronic emergents are also diachronic in the sense that they were unpredictable before their first appearance (Stephan 1998). However, the reverse is not true, that is, diachronic emergents are not necessarily synchronic as well.

Diachronic and synchronic determinations are interwoven in Peirce's philosophy. Semiosis produces novel and unexpected results over time, while also, at all times doing so under the influence of a self-organizing principle which is irreducible to component-relations. This joint action of diachronic and synchronic determinations underlies the notion that semiosis is temporally distributed: at any momentary occurrence it mediates a general temporal development; it subsumes a-temporal feelings and reactions under a continuous, dynamic, cumulative, and self-organized development. This conflation of synchronic and diachronic can be found throughout Peirce's philosophical system, from the growth of habits to the classes of

[3] Peirce's concept of a diagram holds a central position: abduction generates a comprehensive hypothesis with a diagrammatic nature; deduction engages in the manipulation of this diagram; induction aims to empirically verify these diagram-based findings against real-world facts.

immediate, dynamical, and final interpretants, all of which embody the notion of development and growth within an emergent context.

Semiosis is a process that diachronically produces novel and unpredictable outcomes. Dynamic sign-object interactions produce dynamic interpretants, inquiry produces beliefs, and, more broadly, feelings and reactive interactions produce habits. In all of these cases, predictability is limited, and the determination of outcomes is non-linear. Notions such as Firstness, iconicity and abductive inference underlie the notion that vagueness is an inherent component in semiosis, even in the emergence of beliefs. Abductive inference, the first phase of inquiry, may amount to pure guessing. Abduction fashions hypothetical explanations for surprising phenomena, which are then developed deductively and tested inductively. Through an accumulation of new auxiliary hypotheses, deducted consequences, and inductive evidence, a belief grows. However, as synechism emphasizes, this growth remains open to continuous development, and is thus both made possible and bound by the possibility of new surprising phenomena (Hausman 1991).

Whether in diachronic interpretant production or in the growth of beliefs, a similar story is told, involving the categories of Firstness, Secondness, and Thirdness. Thirdness, as in symbols, habits and beliefs, is a category of stable and reproducible structures, which nevertheless are not given in advance, and must spring forth from a temporal development that organizes Firstness (icons, hypotheses, vague possibilities) and secondness (indexes, deductions, determinative compulsion) into a logically stable form that carries the conditions for its own further reproduction, without, however replacing vagueness for absolute regularity, which would amount to pinning down a phenomenon to ultimate exactitude, breaking synechistic continuity.

At the same time that it emphasizes diachronic development, Peirce's system of categories assumes a synchronic hierarchy of presupposition and containment of relations: monadic relations (Firstness, signs-in-itself) are contained and presupposed by dyadic relations (Secondness, sign-object), and these by triadic relations (Thirdness, sign-object-interpretant). The sign-object-interpretant relation is thus not a flat relation involving three elements, but a nested organization that can be represented thusly: (S-O-I(S-O(S))). Habits, for instance, are instantiated in countless material relations and their surrounding space of possibilities for action, while at the same adding to them the property of general tendency of development. Rather than being just a product or an ulterior phase in an unfolding process, Thirdness may be considered a systemic property underlying the whole process. From this perspective, thirdness is synchronically determined by secondness and firstness while at the same time downwardly regulating their interaction. While the first appearance of a tendency may be characterized as a process of diachronic determination and transformational emergence, the overwhelming majority of semiotic interactions involve the synchronic determination between firstness, secondness and thirdness.

3.5 Hierarchical or Flat Emergence

There are two fundamental perspectives on how to conceptualize the difference between bases and emergents, one based on synchronic and hierarchical levels, and another based on diachronic and flat domains. The first perspective, often referred to as holistic or hierarchical emergentism, conceptualizes reality as being structured in levels. These levels may be considered ontological, such as in the case of organizational levels, or merely epistemic, such as in the case of observational levels. The notion of level emphasizes relational notions such as organization, structure, and form. In hierarchical emergentism, bases and emergents may share material substance, and yet possess different forms, or organizational properties. This perspective is often associated with the slogan "the whole is greater than the sum of its parts": interconnected components collectively and/or distributively form an emergent level that exhibits properties that are distinct from the base level. Additionally, holistic emergentism may admit the notion of downward causation, which implies that the emergent level has the capacity to influence the properties or causal powers of the constituent elements at the base level.

In contrast, flat, or transformational, emergentism challenges the idea of hierarchical levels and instead proposes that emergence occurs among domains, without any notion of level. Unlike levels, domains do not admit any synchronic co-existence of bases and emergence. Instead, a base domain diachronically transforms into an emergent domain. The differentiation between distinct domains does not revolve around forms, structures, or organization patterns, but instead hinges on transformations on the component elements themselves: an emergent domain (diachronically subsequent) has different elements (therefore irreducible) from a base domain (diachronically antecedent). Here, the slogan "the whole is greater than the sum of its parts" is replaced by "the subsequent is new in relation to the antecedent." A challenge for flat emergentists is to disambiguate between emergence and regular, non-emergent, processes of transformation.

As we will explore in the third part of the book, our approach to emergentism conceptualizes the difference between bases and emergents in terms of niches of semiotic artifacts,[4] which can be roughly defined as signs-in-action materially instantiated and stabilized through their material embodiment. As semiotic

[4] There is a notable incompatibility between the current notion of an artifact as a material substance or structure transformed by human action and the notion of a semiotic artifact. Material artifacts that situate and distribute processes of meaning and cognition are not "things" but semiotic processes, signs-in-action. Language, observation and measurement instruments, maps and diagrams, musical notations, scientific and mathematical models, algorithms and computers, as well as bodily extensions and couplings such as glasses, are signs-in-action (emergent habits), stabilized and instantiated in physical structures. As signs-in-action, they are situated in material and temporal contexts as producers of regular effects within these contexts. Niches of semiotic artifacts are structured spaces with fundamental conditions for the stability of semiosis, such as situationality (co-location) and temporal distribution among communities of agents, artifacts, and their environments. Niche of semiotic artifacts provides conditions for the emergence of habit and surprise in semiosis/cognition.

emergence unfolds, it sees the emergence of new niches. Unlike the flat domains posited by transformational emergentists, niches do entail multilevel relations. An artifact considered in its intrinsic properties only amounts to a set of monadic relations, which can only be vague. Artifacts only acquire and express the capacity for actually producing interpretants when they are coupled to dynamic environments of constraints and tendencies. This necessarily involves multilevel relations among the inherent embodied properties of an artifact, the situated properties that emerge in the relation between artifact and its local environment, and the distributed properties that emerge from the relation between the situated artifact and general habits of interpretant production. Therefore, niches are better characterized as emergent hierarchical levels than emergent flat domains. However, this does not rule out the possibility that niches, being hierarchical multilevel structures, transform into new niches by a process that is not due to organizational changes in its bases, but rather by material transformation of its bases. We are considering here, as is the case with synchronic and diachronic determinations, that the synchronic/multilevel perspective is more permissive and includes the diachronic/transformational perspective.

3.6 Epistemological or Ontological Emergence

Sartenaer (2016) proposes a taxonomy of emergentism, based on available strategies to balance monism and pluralism. Monism and pluralism can be about substances, properties, and/or predicates and propositions. Emergence theories combine monism and pluralism in different ways, but complete monism or pluralism about all three are not considered emergentist positions. Monism about substance, properties and predicates is *reductive physicalism*,[5] and pluralism about all of them is *substance pluralism*. Emergentism falls between these two extremes and can take two forms: *representational, or epistemological, emergentism*, which involves monism about substances and properties but pluralism about predicates, and *causal, or ontological, emergentism*, which involves monism about substances but pluralism about properties and propositions. Representational emergentism entails causal continuity but representational discontinuity, and causal emergentism entails the opposite.[6]

[5] The characterization of radical monism as physicalism, and not, say, as subjective idealism, presupposes a materialist and physicalist ontology as more scientifically sound.

[6] Sartenaer also identifies different subtypes of representational emergence, namely, theoretical discontinuity, where there are theoretical propositions that cannot be logically derived from substances/properties, and explanatory discontinuity, where predicates cannot be explained on the basis of substances/properties.The analysis yields three main types of emergentism: theoretical emergence, characterized by the idea that theories about the whole cannot be logically deduced from theories about the parts; explanatory emergence, characterized by the idea that the behavior of the whole cannot be fully explained solely based on the knowledge of the behavior of the parts themselves; and causal emergence, which is characterized by the idea that the whole exhibits

Peirce's metaphysics is monist, but considers that matter and mind have different temporal properties. It differentiates between matter and mind, but that differentiation is not a dualist separation of substances (mind is made of something and matter is made of something else) or of localities (mind is "here", matter is "there"). Synechism opposes substantial dualism—the notion that psychical and physical phenomena are two completely separated categories of being—which Peirce refers to as "the philosophy which performs its analyses with an axe, leaving, as the ultimate elements, unrelated chunks of being" (Peirce 1893/1998, EP 2: 2). At the same time, synechism also opposes reducing the psychical to the physical or the physical to the psychical:

> [M]aterialism is the doctrine that matter is everything, idealism the doctrine that ideas are everything, dualism the philosophy which splits everything in two. In like manner, I have proposed to make synechism mean the tendency to regard everything as continuous. (EP 2:1)

Contrary to dualism, Peirce proposes that materiality is "effete mind" (CP 6.25, see also CP 7.364). The distinction between matter and mind, here, is softer, and can be understood in terms of a temporality development. Purely mechanical events—dyadic events—can be described merely in terms of an ever-updated present. Mental processes, on the other hand, need to include a sense of interweaving future and past: the development of a general idea throughout a trajectory. In triadic relations, what happens here-and-now is influenced by a generalized past and an anticipated future. Differently from mechanical actions, minds act towards future outcomes (CP 1.269). This future-orientedness is related to the tendency of sign action to generate more of itself (anticipating what is going to happen in the future). In a triadic relation, something stands for something else only insofar as it produces that relation again, as a future effect. If mental action is temporally distributed, the materiality of mental action is a "frozen" picture of that distributed trajectory of action. For example, a book materializes (embodies into physical regularity) an anticipation or prediction of the reader's interpretative behavior. A book is a temporally distributed activity that involves a generalized past and predicted future tendency. A material copy of the book is a "frozen" picture of this temporally distributed activity, a mediator, for a present experience (the reading event), of general past and an anticipated future.

There is no strict separation, here, between mental processes and material processes: "as far as the process of nature is intelligible, so far is the process of nature identical with the process of reason; the law of being and the law of thought must be practically assumed to be one" (MS 956, see also discussion of this passage in Tiercelin 1998, p. 11).

---

entirely new causal powers that are not identical to any combination of the causal powers of the parts (Sartenaer 2016, p. 11).

3.7 Synechism: Continuity in the Semiotic Process

Synechism can be compared to emergence, firstly in the sense that it is based on a kind of non-reducibility, and secondly in the sense that it conceives of an interconnected process that cannot be broken down into disjointed parts. Synechism means that all parts of the semiotic process are joined together, and this juncture has the character of an infinity greater than and irreducible to any infinite collection of particulars. The semiotic process is so complexly interwoven, that this interweaving defeats any attempt of reductive generalization from a collection of observations. It is important to stress, here, that generalizations, for Peirce, cannot reach a true continuity. Generalities are rudimentary forms of continuity (CP 6.172).

In the Peircean emergentism we propose, mereology is fundamentally temporal. That is, the process is hierarchical and contains components, but these components cannot be fully observed in any synchronous slice or window. In a synchronous slice, what is observed are transformational determinations. It is only from a temporally distributed, historical perspective that the mereology of the process can be observed. This is expressed, in Peirce's philosophy, in the notion of triadic irreducibility and the final interpretant—the process is irreducibly triadic insofar as it genuinely depends on a final interpretant. However, this final interpretant is a mode of being that is temporally distributed.

Chapter 4
The Emergence of Sign Action: The Case of Classical Ballet

Peirce's pragmatic notion of semiosis can be described in terms of a multi-level system of constraints involving chance, efficient, formal and final causation. According to the model proposed here, law-like regularities, which work as boundary conditions or organizational principles, have a downward effect on the spatio-temporal distribution of lower-level semiotic items. We treat this downward determinative influence as a propensity relation: if some lower-level entities a, b, c,…, n are under the influence of a general organizational principle, W, they will show a tendency to behave in certain specific ways, and, thus, to instantiate a set of specific processes. Our goal in this chapter is to examine the role of downward determination in semiotic systems, conceived as multi-level hierarchical systems.

4.1 Classic Ballet Emerges

In this chapter, we explore C. S. Peirce's pragmatic conception of sign action as a temporally distributed and emergent process. We illustrate our argument through the emergence of classical ballet as a semiotic process. Pragmatism was originally formulated by Peirce as a theory of meaning, a methodological principle or rule to clarify ideas, concepts and propositions (Hookway 2002; Legg and Hookway 2019). This principle states that the meaning of a sign implies likely events to occur in a course of experience, according to a set of antecedent conditions (CP 5.468). In this formulation, meaning is an open ended process, constrained by likelihoods, but open to novelty. The meaning of a sign is not a given content nor an essential form, but emerges as a habit of sign action.

The notion of habit occupies a central position in Peirce's pragmatism (Hookway 1985; West and Anderson 2016). Habit entails a disposition to act in certain ways under certain circumstances, especially when the carrier of the habit is stimulated,

P. Atã, J. Queiroz, *Mind as an External Sign: A Semiotic Externalist Philosophy of Cognitive Science*, Synthese Library 531,
https://doi.org/10.1007/978-3-032-08988-5_4

animated, or guided by certain motives (CP 5.480). The meaning of a Peircean sign is most adequately understood through the habits of action they provoke, sustain, and modify. Semiosis emerges (see Queiroz and El-Hani 2006a, b). The emergence of a sign in action is the emergence of an autonomous, or self-organized, habit. Self-organization is a process that leads a complex system towards dynamically stable states of increased redundancy and reduced variability. In our description, stability is not absence of change (stasis), but regularity through change (Kelso 1995). New habits of sign action emerge from previously existing sign action in a cultural evolutionary process. Out of semiosis that follows some stability and tendency of development, new habits become stabilized and begin to develop autonomously. In our description, this corresponds to the emergence of a distributed cognitive system of agents and artifacts which both enacts this habit and is regulated by it (Atã and Queiroz in press).

An example is the historical emergence of classical ballet. Classical ballet is a habit of sign action: a codification of dance steps that disciplines bodies and stabilize their communicative behavior within a community of utterers and interpreters. Any ballet performance is an instantiated sign, composed of several signs, and which determines interpretive effects according to some regularity. In any ballet performance, a sign acts according to a regular habit that connects sign (performance), object (in a most general description, the 'form' of ballet itself that regulates how signs behave), and interpretants (interpretive effects, for example on the audience).

The historical emergence of classical ballet is a self-regulatory process, in which a system of different kinds of cognitive artifacts (musical, bodily/motor, visual, spatial/architectonic) and agents obtained a stable semiotic relation throughout many phases of development between the sixteenth and nineteenth Centuries. From the sixteenth to the nineteenth centuries, the ballet sign has taken several habits that regulate sign action, such as the codification of a vocabulary of dance steps with a specific nomenclature and notation, the establishment of the proscenium arch stage as the preferred technology regulating performance space, the standardization of training techniques and consequent relative regularization of the skills, bodies, and abilities of professional dancers, etc. All of these are cases in which variability has been decreased and redundancy increased, so that sign action becomes more regular. They all are part of what makes classical ballet a sign in action, endowed with a potential to further produce interpretants. This is not only emergence of actual meaning, but also, emergence of an open-ended niche of potential and general meaning processes, an autonomous tendency of development or a pragmatic likelihood of occurrences.

4.2 Semiosis Is Temporally Distributed

Semiosis is a dynamic process, with signs continuously translating into other signs, in time. The mode of existence of a sign is to produce interpretative effects which are new signs. A single person thinking to herself already embodies a dynamic process of semiosis. This temporal distribution of semiosis is a central aspect of Peirce's semiotics: "To say, therefore, that thought cannot happen in an instant, but requires a time, is but another way of saying that every thought must be interpreted in another, or that all thought is in signs" (CP 5.253). Semiosis is temporally distributed, a time-dependent action of signs which develops in several historical and evolutionary scales. To describe a habit of sign action is to describe a dynamic process involving past and future: a relation, developing in time, between a cumulative generalization of a past history of sign actions and a likelihood that this history will somehow regulate future sign actions. In this framework, stability of semiosis in time is not understood as an instantiation of representational states in a system but as regularity of sign action throughout communication. A habit doesn't entail stasis, but regularity throughout change.

Classical ballet entails semiotic chains in several different time scales. An instantiated sign of ballet, such as a ballet performance, has the power of generating new signs: a ballet performance generates interpretative effects in an audience, which develop into a critic literature about that performance, which in turn generates more interpretive effects, and so forth, constituting traditions of critique and discourse as sequences of reinterpretations and retranslations of ballet signs. Ballet performances also generate all sorts of artworks in different media (including other ballet performances) that dialogue with it by making reference, reinterpreting themes, taking inspiration, reacting aesthetically, and so on. But we don't need to consider only instantiated ballet performances as examples of how ballet entails semiotic chains. The body of the ballet dancer is a sign, being reinterpreted and translated throughout the history of ballet. The baroque courtier dancer of the seventeenth century is a sign endowed with a potential to generate interpretants. It gets translated into the professional academic dancers, the dancers of different ballet schools, all the way to modern and contemporary dancers. What connects all of these dancing bodies is a temporally distributed semiotic process, a semiotic chain that communicates some balletic forms so as to generate new ballet signs.

This process also entails an interplay between redundancy and novelty, a negotiation of the stability of semiotic habits. One of the most stable habits of classical ballet is the codification of the so-called five basic positions of the body (or, more precisely, the feet). The first position is a 'gathering point' or resting position for the dancer (heels meeting at a central point and toes pointing outward), while the other four positions prepare the dancer to move (Homans 2010, p. 23) (see Fig. 4.1). Codified in the late seventeenth century by Pierre Beauchamp, the dancing master of Louis XIV, the five positions are the foundations of classical ballet:

The importance of these positions cannot be overstated: they are the major scale, the primary colors from which all other constructions in ballet arise. Without them,

Fig. 4.1 Five basic positions of the body

la belle danse was a social dance; with them, the crucial leap from etiquette to art was made (Homans 2010, p. 23).

Between sixteenth and nineteenth centuries, dance changed radically: baroque dance became classical ballet, the nobles and courtiers were replaced by professional dancers, the palace halls and gardens were replaced by theaters with proscenium arch stages, the male predominance on stage was overcome by the central figure of the ballerina, and yet the five positions changed very little in comparison. To this day, they remain structurally the same, with the main transformation being a widening of the angle at which the feet point outward from the body (originally, the feet were never supposed to point outwards at an angle wider than 45°).

However, this semiotic stability is not the same as propagation of a representational state or of informational content. The five positions of the feet discipline the dancer's movement so as to constrain the interpretative effects that the dance performance entails. Originally, this codification was carried out with the objective of making the dancer appear noble, as opposed to lesser social characters that were depicted as having the feet pointing inwards (Homans 2010, p. 23). But as ballet became more detached from etiquette, this distinction between the noble body and the body of the lower social classes became less pragmatically relevant for the action of the sign. In our description, the five positions of the body cannot be said to be a representational state or to convey an informational content. Its stability has the form of a habit of action, with open-ended results, across very distinct contexts and communicative situations. The semiotic character of the five position of the feet is historical, temporally distributed: it is not synchronically located, and its object is a mere form that requires time and context to be expressed as a regularity.

This dynamic semiotic frame needs to be understood together with Peirce's pragmatism. The mode of being of signs is to act—determine effects—in the world. Semiosis is a process in which action and communication are co-dependent. This entails a semiotic kind of cognitive externalism: just as for Peirce it is impossible to think without signs, so also cognition itself is impossible without external, materially-incorporated action and its consequences. Minds are not severed from the external world, but are continuous to it. To think with signs is to think with external artifacts and within a community.

This view of the temporal distribution of meaning has consequences for the notion of distributed cognitive systems (Zhang and Patel 2006; Hutchins 2014; Davies and Michaelian 2016), as it entails a 'semiotization' of the discussion about distribution of cognition (see Atã and Queiroz in press). Hutchins (2014) has characterized distributed cognition as an emergentist approach to cognition. In our approach, semiosis is a temporally distributed process in which a regular tendency towards certain future outcomes emerges out of a history of sign actions (a semiotic chain). Our approach here is pragmatic and action-oriented (cf. Engel et al. 2016; Madzia and Jung 2016), and focuses on sign action, as conceived by Peirce's pragmatism. According to the kind of cognitive externalism afforded by a Peircean perspective, whenever we consider distributed cognitive systems, we have to consider that the agents and artifacts in question embody irreducible semiotic processes. What makes some entity a part of a DCS is that, in the context of that system, the entity acts as a first term that triadically stands for a second term in order to determine a third term, which is a regular effect. In other words, what makes an entity a part of a DCS is that it acts as a sign so as to generate other signs. Why is it crucial to consider semiosis when describing DCSs? We cannot explain (it is not sufficient) the meaningful behavior of DCSs as the effects of explanatory units such as persons, groups of persons, or artifacts. The phenomenon to be described and explained when examining a DCS is the action of signs. Agents and artefacts in a DCS are observed as processes of irreducible triadic relations. DCSs and their meaningful behavior are outcomes of the action of signs. The fundamental unit of explanation for the behavior of persons and artefacts in a DCS are the signs that regulate them. The concept of semiosis is not a re-description of individual actions of agents and artifacts. In fact, the object of description here is not agents, artifacts, and their doings, but some real general process that supervenes on them but is not reducible to their particular properties. In this view, whenever one refers to an agent or a cognitive artifact in a DCS, he/she is referring to signs, and describing semiosis. In this active semiotic externalism, cognition is the development of available cognitive artifacts in which it is embodied as a power to produce interpretants (interpretative/communicative effects).

Consider ballet: It is a habit of sign action: a codification of dance steps that disciplines bodies and stabilize their communicative behavior within a community of utterers and interpreters. This habit is temporally distributed. In the case of ballet, the agents of the DCS are dancers, audiences, choreographers, producers, theorists, critics, etc., that integrate a community within which 'ballet' is a sign endowed with meaning. The artifacts of the DCS integrate an extensive list of items with which this large community enact the sign 'ballet' as a process of meaning: dance notation systems, markings, musical notations and instruments, stage architecture and stage technologies, illumination devices, wearables, props and prostheses, literary artifacts such as narrative structures and character archetypes, print media such newspapers, magazines, books, etc. This extensive distributed cognitive system develops in time, from sixteenth century court dance to twentieth century classical ballet. Throughout this time, agents and artifacts come and go, but the distributed system continues as a dynamical regularity throughout radically different circumstances and situations. This dynamical regularity evolves, changing its conditions of stability in several critical moments. This evolution takes the form of development of cognitive artifacts.

4.3 Emergence of Semiosis

Semiosis is an emergent self-organizing process in a complex system of sign users interacting locally and mutually affecting each other (Loula et al. 2010). Inspired by Stanley Salthe's (1985) hierarchical structuralist model of emergence, Queiroz and El-Hani (2006a, b, 2012) have modeled semiosis as an emergent process. A fundamental element in Salthe's hierarchical structuralism is the "basic triadic system", clearly influenced by Peirce's theory of categories. According to the basic triadic system, to describe the fundamental interactions of a given entity or process in a hierarchy, we need (i) to consider it at the level where we actually observe it ("focal level"); (ii) to investigate it in terms of its relations with the parts described at the next lower level; and (iii) to take into account entities or processes at the next higher level, in which the entities or processes observed at the focal level are embedded. In Salthe's triadic system, both the lower and the higher level have constraining influences over the dynamics of the entities and/or processes at the focal level. These constraints allow us to explain the emergence of entities or processes (e.g., semiosis) at the focal level.

In this model, an emergent process at the focal level is explained as the product of an interaction between processes taking place at the next lower and higher levels. The phenomena observed at the focal level should be "… among the possibilities engendered by permutations of possible initiating conditions established at the next lower level" (Salthe 1985, p. 101). Nevertheless, processes at the focal level are embedded in a higher-level environment that plays a role as important as that of the lower level and its initiating conditions. Through the temporal evolution of the systems at the focal level, this environment or context selects among the states potentially engendered by the components those that will be effectively actualized. As Salthe (id. ibid.) puts it, "what actually will emerge will be guided by combinations of boundary conditions imposed by the next higher level".

Following this model, the description of semiosis needs to consider at least three distinct levels of description: a focal level, a micro-semiotic level, and a macro-semiotic level. The idea here is that an observer (such as a researcher doing analysis) can set a focal level of observation of the semiotic process, and from that perspective identify a level which is lower (a micro-level, more fine-grained) and a level which is higher (a macro-level, more encompassing).

The focal level is the level of observation of an actual phenomenon. This level is comparable to the Peircean logical phenomenological category of secondness, the consideration of actuality of situated experience. At this level we observe signs effectively acting and producing new signs. This dynamic action of signs has been modeled by Peirce as a chain of semiotic triads. This effective sign action is emergent, which in our approach means that it cannot be logically inferred from the behavior of its component parts considered in simpler relations.

The micro-semiotic level is a lower level of description, which considers the components of the focal phenomenon. These components could be related in many possible ways, and the actual semiosis observed at the focal level is but one among many ways the phenomenon could have been instantiated. This level is comparable to the Peircean logical phenomenological category of firstness, the consideration of elements of situated experience regardless of any actual connection or determination. At this level of description we can only regard the elements in their positive characters and situated possibilities of action, how they could act, but not how they will or would act. Here we observe not an actual semiotic chain, but only possible signs determining possible interpretants in relation to possible objects. The examination of a micro-semiotic level of description establishes initiating conditions, or potentialities of action, but how a sign will act is indeterminable by this level alone. Hence, besides considering the micro-semiotic level of description we need also to consider a macro-semiotic level.

The macro-semiotic level is a higher level of description, which considers an environment or context in which the focal phenomenon is framed or niched. This level is comparable to the Peircean logical phenomenological category of thirdness, the consideration of elements of situated experience as they are mediated by habits (general tendencies, regularities or law-like behaviors). In our description, this contextual level is a temporal context, and the habit is temporally distributed in the form of a cumulative and generalizable history of past sign action that tends towards some futures rather than others. This tendency plays the selective role of boundary conditions which limit or constrain potentialities of action in terms of historical tendencies. At this level we describe a history of interconnected semiotic chains forming a past semiotic web or network that embeds habits of sign action. But the examination of a macro level of description doesn't allow us to infer how these habits will be instantiated. In order to do that, we have to consider the interaction between micro-semiotic and macro-semiotic level.

The emergence of effective sign action at the focal level is the result of an irreducible interplay of initiating and boundary conditions. An upward (micro to focal) constitutive determination on what will emerge can be described in the form of a potentiality of sign action provided by the components of a semiotic triad: potential signs, potential objects, potential interpretants. A downward (macro to focal) selective determination on what will emerge can be described in the form of a tendency of sign action given by a temporal or historical context. The irreducible interplay of these upward and downward determinative relations result in the emergence of effective sign action observed in the focal level (see Figs. 4.2 and 4.3).

An important assumption of this model is the ubiquity of semiotic emergence. In semiotic systems, every sign action is emergent. This is compatible with Peirce's distinction between dyadic brute force interactions on the one hand and mediated, semiotic, interactions on the other. A consequence of this view is that semiotic chains and networks are constantly subjected to potential emergence of novelty. The

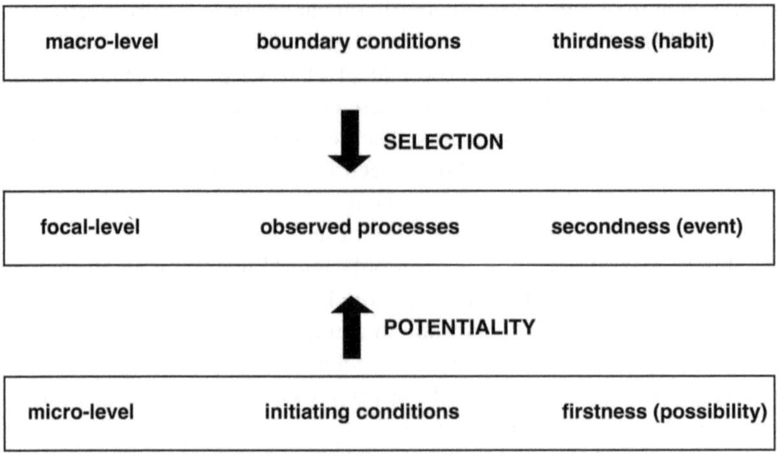

Fig. 4.2 A scheme of the determinative relationships in Salthe's hierarchical system in relation to Peirce's categories. Initiating conditions (associated with Peirce's category of firstness) and boundary conditions (associated with Peirce's category of thirdness) act together and irreducibly, determining an upward potentiality of action and a downward selective determination on the emergence of observed processes in the focal level (associated with Peirce's category of secondness)

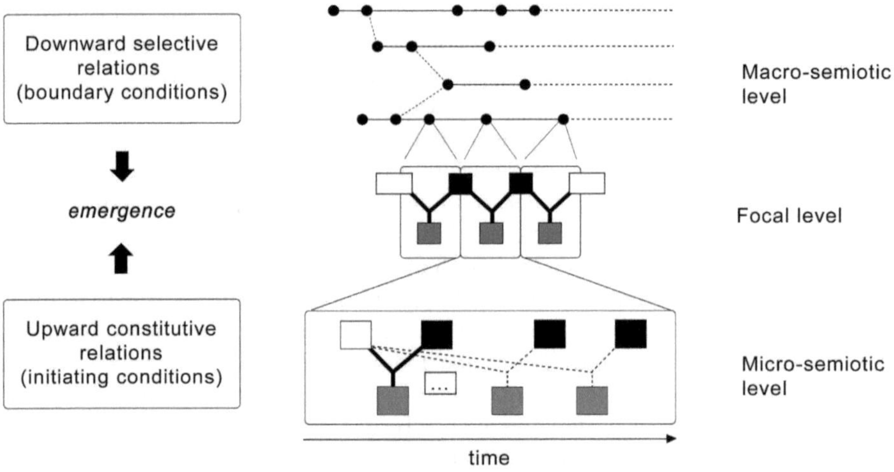

Fig. 4.3 Micro-semiotic, macro-semiotic, and focal level of observations. At the focal level we observe effective dynamic sign action that can be modelled as semiotic chains. At the micro-semiotic level we observe possible sign action (represented by the dotted lines between possible semiotic triads). At the macro-semiotic level we observe general sign action, in the form of networks of semiotic chains regulated by habits. The interaction between the micro-semiotic and the macro-semiotic level leads to the emergence of actual semiosis at the focal level

stability of semiosis in time is not conceived as absence of novelty, but as regularity obtained across novelty.

Another important assumption of the model is the self-organization of semiotic chains and networks. Semiotic networks at the macro-semiotic level are constituted by past semiotic chains at the focal level, and in turn, help to further determine it. A consequence of self-organization of semiosis is that sign action is always to some degree self-referential: ballet signs instantiate ballet objects, ballet properties, ballet experiments, etc., that are considered pragmatically relevant in a certain context of practice that has itself been constructed by these objects, properties, experiments. This view is in opposition to the understanding of signs as conveyors of referential content. It is simplistic to attempt to characterize the semiotic process of ballet as dependent on some referential object independent of the semiotic process itself, because the whole complex is irreducibly dependent and self-constructed. The pragmatically relevant communication of self-referential properties of ballet renews or reiterates the capacity of this sign or system of signs (ballet) of semiotically structuring some contextual reality, or of conferring some useful agency to utterers and interpreters in relation to a given contextual reality. In this sense, whenever we talk about dancers, audience, or any cognitive artifacts involved in the action of the ballet sign, we consider them as signs whose action is emergent and integrate a self-constructed semiotic process.

4.4 Emergence of Classical Ballet as a Sign in Action

Classical ballet is an emergent sign in action. This sign in action is distributed in time (across the whole history of ballet), and defined as an open-ended tendency of development and pragmatic likelihood of occurrences of meaning processes. The emergence of classical ballet is the emergence of a habitual, self-organized, temporally distributed semiotic process. It involves several historical episodes, such as the leap from etiquette to art form with the codification of body positions during the reign of Louis XIV, the transfer of dance performances from court venues to theaters with proscenium arch stages, the professionalization of dance and the formation of professional dancers, the formation and development of a popular audience for ballet, several episodes of embodied research and stabilization of dance techniques (e.g., the emergence of the pointe technique, popularized by Marie Taglioni in the 1830s), several episodes of systematization and codification of dance vocabularies (e.g., such as the alphabet of dance steps published in Carlo Blasis' 1828 The Code of Terpsichore). At any of these episodes, the development of the semiotic chain is subjected to upward initiating conditions from a micro-semiotic level and downward boundary conditions from a macro-semiotic level.

A particular example that can illustrate the role of micro-semiotic and macro-semiotic levels concerns the proscenium arch stage. At a micro-semiotic level, the development of the semiotic chain of classical ballet is regulated by the availability of artifacts and the initiating conditions for sign action they embody. Consider the performance space of ballet. In its baroque origins in European courts, ballets were performed in palace halls or gardens, there were no elevated stages or proscenium arches, and the audience occupied seats arranged in tiers and viewed the spectacle from above (Cohen 1992; Homans 2010). This positioning of dancers in relation to the audience afforded geometric patterns of dancers' displacement in the performance space:

> [T]he ballets were performed, not on raised stages, but in the central space of large halls with the audience seated above the floor in galleries that extended around three sides of the dancing area. The wise masters reasoned out that the way to dazzle was not with steps, which the performers could not do expertly and the audience could not see well, but with floor patterns — complex geometrical shapes that formed, dissolved, and reformed to display a tantalizing variety of designs. (Cohen 1992, p. 7)

The historical development of classical ballet saw a transition from these performance spaces to proscenium arch stages as we typically know them. That corresponded to a change in the relation between dancers and audiences. The architecture of the proscenium arch stage offers a different set of initiating conditions for sign action incorporated in the performance space. as it locates the performance in front of the audience and frames it with the proscenium arch. The development of this kind of stage was directly influenced by visual perspective (see Breton 1990; Barker and Bay 2016), a technique developed by Filippo Brunelleschi and Leon Battista Alberti in the context of architectural drawing and painting. The stage functions as a one-point perspective box that privileges frontality of observation.

Regardless of wider historical or contextual considerations, the frontality of observation afforded by the cognitive artifact of the proscenium arch stage already imposes new initiating conditions for sign action. However, when we consider actual semiosis, this set of initiating conditions are always acting together with a macro-semiotic level, subjected to a regulatory process that involves a network of semiotic chains. When we consider the proscenium arch stage in ballet, a relevant semiotic chain concerns the construction of the body of the dancer as a sign endowed with meaning. This extensive semiotic chain includes, among others, the contrapposto of Classical Antiquity statuary, the Renaissance aesthetic ideal of mathematically harmonious body proportions, the Pythagorean academic association between body, music and mathematics which was popular in sixteenth century France (Homans 2010, p. 5), the severe discipline of body movements established by the etiquette of the French court, the posture and footwork of fencing. This historical regulation had before determined the geometrically-precise body of the baroque dancer as the dancing body performing in the halls and gardens for an audience of nobles and courtiers. The introduction of the proscenium arch stage and its new set of initiating conditions was followed by new retranslations of the ballet dancer as a sign. In the eighteenth century, the ballet master creator of the ballet d'action, Jean George Noverre, likened the perspective box of the proscenium arch stage to the tableaux of a painting, and put forward a conception of ballet as a dynamic painting. Noverre emphasized pictorial composition and narrative, chiaroscuro, and perspective, suggesting that dancers be assigned roles according to their stature so as to emphasize the perspectival illusion of depth, and introducing pauses in the ballet action so that the audience could take time to appreciate the details of the visual compositions of his balletic tableaux (Monteiro 1998, Chap. 2; Homans 2010, p. 74). The intersemiotic relation (relation between different semiotic systems) with painting regulated how the frontality of observation afforded by the proscenium arch stage theater acted semiotically in the ballet d'action. The frontality of observation afforded by the proscenium arch stage also provides initiating conditions for the romantic revolution that marks the start of modern ballets (see Homans 2010, p. 170). Romantic ballet was decisively dependent on vertical morphologies of dance movement. The influential dancer Marie Taglioni, the first romantic ballerina, stabilized into a balletic habit the en pointe technique (dancing on the tip of the toes), helping to form the ideal image of the ballerina as an ethereal and weightless figure. This verticality of dance went together with Romanticist aesthetic ideals. In fact, as indicated by Homans, Taglioni's career represents a central point of connection between ballet and French literary Romanticism:

> Robert le Diable [1831 opera including a ballet performed by Taglioni] opened ballet to the world of literary Romanticism. In the years to come, a generation of poets, writers, and artists found themselves drawn to Taglioni and to dance. Heine, Stendhal, Balzac, Théophile Gautier, and Jules Janin all wrote about ballet. The poems and stories of Sir Walter Scott and E. T. A. Hoffmann, Victor Hugo and Charles Nodier, served as inspiration for ballet masters, and both Heine and Gautier wrote ballet librettos of their own. Perhaps most important of all, and building on Noverre, these poets and writers understood that ballet was not merely an aspect of opera but had a distinct language of its own: they were its first informed critics. Nor was their role merely responsive or passive, for their writings defined Taglioni's image and played a critical role in promoting her career. (Homans 2010, pp. 150–151)

Both Noverre's pictoric ballet and the verticality of dance of the romantic ballerina are retranslations of the body of the ballet dancer as a sign. They presuppose the frontality of observation afforded by the proscenium arch stage, and submit it to the regulatory influence of past sign action (perspective in painting in the case of Noverre, literary Romanticism in the case of Taglioni). These operations create novel processes of meaning, emergent episodes of sign action.

4.5 Temporal Dynamics of Distributed Sign Action

In classical examples of distributed cognition, sign action is distributed within spatial contexts: boats (Hutchins 1995a), cockpit of an airplane (Hutchins 1995b), research laboratory (Nersessian et al. 2003), theater building (Tribble 2005). In these examples, the distribution of cognition is decisively conceived as a matter of spatial locus: cognition is not only located in the head of individuals, but in spatial environments of cognitive artifacts and cultural practices. In our approach, the focus of description of distributed sign action is not spatial, but temporal. It is not only the case that a sign cannot simply be described as spatially contained within one's skull, it is also (and more centrally) the case that a sign cannot be described as temporally contained within one event or episode. Even when the focus of observation is effective instantiated sign action in a single event, the sign acts as a temporally distributed process in which a regular tendency towards certain future outcomes emerges out of a history of sign interactions.

Our most fundamental explanatory unit is sign action, a temporally distributed process. This is what grounds our emergentist account of cognition. Semiosis self-organizes in time, in a process that continuously entails the production of more signs. Emergence is a ubiquitous condition in this process: the translation of signs into signs cannot be inferred from the properties of the components of a semiotic triad alone, but has to take into account a complex interaction between a micro-semiotic and macro-semiotic level of description. This interaction can be understood as an interplay of potentialities and tendencies, or upward constitutive determinative relations and downward selective determinative relations. According to this view, emergence is a central defining condition of processes of meaning. The emergence of sign action is associated with novelty, creativity, and surprise in cultural evolution.

Ballet is a sign in action. It is a unique process of meaning that continues to emerge in time through the interplay of material situated conditions afforded by cognitive artifacts at a given moment and habits of sign action within networks of semiotic chains. This process has emerged as a relation of meaning involving different kinds of cognitive artifacts: musical, bodily / motor, visual and pictoric, spatial and architectonic, etc. Before the emergence, these cognitive artifacts developed according to various semiotic habits: e.g., etiquette and fencing regulated body discipline, perspective regulated the architectonic space of theatre stages and the pictoric space of the canvas. In classical ballet, a novel and unique habit emerges that regulates the behavior of several cognitive artifacts. One case is the development of the verticality of dance in classical ballet as a semiotic relation connecting proscenium arch stages and dancers as temporally distributed signs. This development is micro-semiotically determined by the spatial conditions of the proscenium arch stage, and macro-semiotically determined by a historical construction of the body of the dancer within a network of semiotic chains, such as the body of the dancer as a pictoric object in a painting tableaux organized by one-point perspective, or as an embodiment of the aesthetic ideals of literary Romanticism. This is not only the emergence of actual meaning, but also the emergence of an open-ended field of potential and general meanings, a self-organized and temporally-distributed semiotic process.

Chapter 5
Artifacts in Problem Solving: Iconicity and Abductive Inference

5.1 Artifacts in Action

Semiosis is materially extended as a process of artifact generation and transformation. Thought and meaning take place in situated environments as the use, development, and transformation of all sorts of cognitive technology: written and spoken language, computer software, maps, graphs and diagrams, plans and blueprints, notations, symbols, numerals and codes, scientific instruments, artworks, bodily expression, and so on. What is decisive, in all of these cases, is that artifacts embody a certain processual structure: they are used to direct action towards anticipated goals or ends. In turn, these goals or ends are also expressed as certain artifactual configurations: science produces theories and models, art produces artworks, thought produces beliefs, and so on. Every meaningful action or process starts and ends with artifacts. In Peirce's terminology, artifacts are Firsts of a semiotic relation, they are signs, which produce interpretants, which in turn are themselves signs for further production of interpretants, in a recursive process of artifact growth.

Problem solving theory models cognition as tasks that involve a problem, a set of rules and a solution set for how to reach a goal while following the rules. As emphasized by the perspective of situated problem solving, this is a process of artifacts, through and through. It is an artifact that embodies a problem in a situated environment: it frames the problem somehow, provides an initial state and communicates a set of rules. It is through the manipulation of artifacts that the problem is explored in search of a solution. Reaching a solution, in turn, is to arrive at a desired artifact configuration, as allowed by the rules.

In this chapter, we are interested in how artifacts transform the space of possibilities in problem solving. In problem solving, artifacts embody a problem and allow work towards a solution. This is a process of meaning: an artifact represents a problem and kickstarts problem solving. It is only through artifacts that an agent can engage with a problem, but there are different ways in which artifacts may embody

P. Atã, J. Queiroz, *Mind as an External Sign: A Semiotic Externalist Philosophy of Cognitive Science*, Synthese Library 531,
https://doi.org/10.1007/978-3-032-08988-5_5

problems for a situated agent, and these different ways may significantly influence problem solving. This is the process of problem framing and registration, the process of construction of a problem space, which traditional problem solving theory has considered as secondary, as a mere preliminary stage to the more crucial phase of search of a problem space (Kirsh 1994). It is also known as the representational effect: how a problem is represented may be crucial for its solution (Zhang and Norman 1994). The representational effect underlies the semiotic nature of problem solving as an activity crucially dependent on sign-object-interpretant relations.

We can investigate the role of artifacts in representing problems by comparing different isomorphic representations of the same problem. In this chapter, we take advantage of Zhang and Norman's (1994) study of isomorphs of the Tower of Hanoi puzzle to investigate the iconic and abductive role of artifacts as signs in problem solving.

5.2 Semiotic Artifacts

The word 'artifact' is derived from the Latin words 'arte', the ablative of 'ars' (meaning 'art'), and 'factum', the past participle of 'facere' (meaning 'to make'). According to Webster's *Third New International Dictionary*, an artifact is typically a simple object, such as a tool or an ornament, that exhibits evidence of human workmanship and modification, as distinguished from a natural object. According to the *Oxford English Dictionary*, an artifact (or artefact) is defined as anything that is created through human artistry and workmanship. In the standard definition, artifacts are intentionally made objects created for a specific purpose:

> I take artifacts (in the strict sense of the word) to be physical objects which have been manufactured for a certain purpose or intentionally modified for a certain purpose. Various cultural objects are created by making artifacts; for example, we can say that artists create works of art by making certain artifacts. (Hilpinen 1992, p. 58)

Hilpinen (1995) later extended the concept of artifact to include belief systems. This definition traces back to Aristotle's differentiation between objects that exist by nature and those that exist by human craft (Metaphysics 1033a ff., Nicomachean Ethics 1140a ff., Physics 192b ff.). Objects that exist by craft originated from the craftsperson, specifically in the form of the object as it exists in advance in the maker's mind. The primary concern of both Aristotle and his modern successors is to distinguish artifacts from naturally occurring objects that do not involve any human intervention. However, if the standard definition of artifacts presuppose the notion of an intentional agent, it is important to note that intentional agency is not exclusive to humans, and not a product of high-level evolutionary morphism. Toolmaking has been observed among several taxa in their natural habitats. The definition of an animal artifact, as proposed by James L. Gould (2007, p. 249): "any creation on the part of an animal, using and/or modifying available materials, which is useful to it or its offspring."

Hilpinen (2011) takes a term from Wendell Oswalt (1973) and calls naturally occurring objects that are used intentionally, without any changes, 'naturefacts'—"Objects taken from their natural environment and used as tools or for some other purpose form a bridge between natural objects and artifacts. Such objects are called "naturefacts" ("made by nature"; Oswalt 1973, pp. 14–17; 1976, pp. 17–23;). When a naturefact is modified to improve its usability for some purpose, it becomes a genuine artifact."

We consider as artifacts anything that can be used as a sign. This may include human-made artifacts, animal-made artifacts, machine-made artifacts, and naturefacts. An example of naturefact in semiosis is a sunflower used as a sign of the sun (see CP 2.274). We will heretofore consider the term artifact to include naturefacts as a subcategory, so that any mention to artifact includes naturefacts. In our approach, signs are artifacts. Signs, as first elements of a triadic relation, embody teleonomic, feedback-driven influences (as well purposes, if we are considering human communication and culture). This is most clear if we start by considering legisigns, which are habits that serve the function of further reproducing that habit. A word, for example, serves a communicative function because it is able to replicate further, in another time and place, the habitual triadic connection it holds to its object and interpretant. Sinsigns and qualisigns are instantiations and components of legisigns, and embody a final cause to the extent that they are always caught up in the process of habit development, as operators of this process, whether as stabilizers or destabilizers of habituality. Treating artifacts as signs suggests that their ontology has to account for semiotic properties, such as temporal distribution, future-orientedness, emergence, self-organization, and distributed agency.

In semiotic externalism, external artifacts do not serve merely as input and output of cognitive processes; they are real constitutive parts of cognition. These external resources are considered cognitive artifacts (Hutchins 1999) that aid, support, enhance or improve cognition. Distributed cognition and situated problem solving traditions have described examples of the use of cognitive artifacts in different domains, for example in numerical cognition (Zhang and Wang 2005; Bender and Beller 2017), practicing and learning science (Pande and Chandrasekharan 2017; Nersessian et al. 2003; Nersessian 2008), situated problem solving (Kirsh 2009; Zhang and Norman 1994), joint creative problem solving tasks (Bjørndahl et al. 2014). Cognitive artifacts are pervasive in our cognition: alphabets and notation systems, numbers, graphs and diagrams, specialized scientific instruments, artistic and communicational media, electronic devices, and pharmacological drugs are some examples of external tools we use to aid, support, enhance or improve cognitive abilities. However, maybe the most impactful example of cognitive artifact is language itself. Language is a scaffolding device, deeply ingrained, that radically augments what brains can do. Language is internal but also external. Using external language, such as in speech and writing, we can augment memory, learning, categorization, attention, our capacity to build social relations and institutions, etc.

5.3 Iconicity and Abductive Inference

The fundamental classification of signs (icon, index, symbol) exhibit a characteristic that can serve as an operational criterion for distinguishing different types of signs: the relative dependence of the components—sign, object, and interpretant (S–O–I)—within the triadic relationship. The icon is the type of sign that depends the most on artifacts. Icons directly depict their objects through the artifact's shape, structure, functioning, or material composition. Their representational quality corresponds with what Peirce characterizes as "pure categorical firstness", an inherent quality within the artifacts themselves, regardless of external factors. This internal aspect belongs to the artifacts, in their own right, as sensible objects, and they would possess it in any circumstance, whether or not there is a real-world object they resemble, even if they have never been interpreted as signs (see CP 4.447). In iconicity, the artifact has the power to determine the object for an interpretant: the S-O relation is S-dependent. This is in contrast to indices, in which the S-O relation depends on a real connection between artifact and object, or to symbols, in which this relation is fundamentally dependent on the triadic mediation of the interpretant. To say that a representation is external in respect to some constraints already implies that these constraints are S-dependent, and that we are dealing with iconic semiosis.

Maps, graphs, charts, algebra, syntax, and formalizations of all kinds, should all be recognized as icons. They are all examples of diagrams, icons of relations. As soon as an icon can be considered as consisting of interrelated parts, and since these relations are subject to experimental manipulation governed by laws, we are working with diagrams (Stjernfelt 2007, p. 92). Diagrams are the principal way of acquiring new knowledge about relations. They represent, through the relations between its parts, the relations that constitute the related parts of the object it represents. The object of the diagram is always a relationship, and the related parts of the diagram represent the relationships that constitute the object represented. The prototypical diagram is described as the manipulation of a geometric figure for the observation of a theorem. But the idea is quite general.[1] An example taken from algebra is enlightening: "In fact, every algebraic equation is an icon, since that shows, through their algebraic signs (which are not themselves icons) relations of the quantities involved" (CP 2.282, emphasis added). Indeed, if a sign is observed as a whole consisting of interrelated parts, and these related parts are subject to experimental modification governed by rules, we are operating with a diagram.

Icons are typically defined as signification by resemblance. What this means is that the icon is capable of revealing aspects or qualities of its object. Trying to make the notion of resemblance more technical, we can adopt an operational definition of iconicity:

[1]Arnold (2011, p. 17) draws attention to the fact that Peirce himself suggested that systems of sonorous and verbal signs can produce iconic and diagrammatic representations (CP 3:418). Pharies is another author who emphasizes that the "physical properties" of verbal discourse can be manipulated to reveal important information about its relational object.

> The key of iconicity is not perceived resemblance between the sign and what it signifies but rather the possibility of making new discoveries about the object of a sign through observing features of the sign itself. Thus a mathematical model of a physical system is an iconic representation because its use provides new information about the physical system. This is the distinctive feature and value of iconic representation: a sign resembles its object if, and only if, study of the sign can yield new information about the object. (Hookway 2002, p. 102)

Operationally defined, the icon is the only type of signs that allows discoveries to be made about the object. This definition approximates the icon to abduction, or hypothesis, the inferential step involved in making discoveries. Abduction is one of three irreducible types of inferences in Peirce's model: abduction, deduction and induction. They correspond to three subsequent phases in the process of scientific inquiry (CP 6.469–473). Abduction rises from the observation of a mass of facts that doesn't fit into the habits and expectations of the observer and culminates with the formation and selection of a hypothesis. Deduction develops testable consequences of the previously generated hypothesis. Based on these consequences, induction performs tests to evaluate it.

Abduction is a distinct form of logical inference, though it can often be and is frequently confused with perceptual judgment in extreme cases. The characterization of abduction as the transformation of mass of facts into hypotheses and the first stage of inquiry brings it close to perception (see Hoffmann 2000; Paavola 2005). For Peirce, perception involves an interpretative process (CP 5.181). It is through an inferential-like perceptual judgment that percepts are subsumed under general classes. This perceptual judgment accounts for the transformation of sense data into knowledge applicable to theoretical or practical use. It is subconscious, but if it was subjected to logical analysis, it would present an inferential—abductive—form (CP 5.181). Therefore, "all that makes knowledge applicable comes to us via abduction" (MS 692).

Peirce defines abduction as 'the process of forming explanatory hypotheses' (CP 5.171), the 'only kind of argument which starts a new idea' (CP 2.96). It involves two operations: the selection and formulation of hypotheses for further consideration (CP 6.525). Functioning as an 'act of insight' that 'comes to us like a flash' (CP 5.181), abduction is relevant not only to the creative and aesthetic dimensions of cognition but also to the significance of abductive inference in formal disciplines. According to Peirce, abduction serves as the logical inference through which new knowledge can be acquired: "Abduction consists in studying the facts and devising a theory to explain them." (CP 5.145). It is the "originary" kind of inference, that is, "the only kind of argument which starts a new idea" (CP 2.96). Magnani (2005) introduces the concept of "manipulative abduction" to refer to those cases where the inference depends on the exploration of external resources—it "happens when we are thinking through doing and not only, in a pragmatic sense, about doing" (Magnani 2005, p. 274). This concept embodies a cognitive process that goes beyond mere consideration of actions; it involves active and deliberate engagement with those actions.

According to Paavola (2011), in abduction the iconic character of reasoning is more prominent. Icons, abductions and perceptual judgments all have important similarities between themselves.

> In all of them, some characteristics or phenomena suggest a potential way of interpreting or explaining these characteristics or phenomena and bringing them into some kind of an order. (Paavola 2011, p. 305)

Paavola has referred to these characteristics that only suggest a way in which they could be interpreted as clue-like characteristics. In abduction, these clue-like characteristics, together with background knowledge, lead to the conclusion of a hypothesis (i.e., a promising way of ordering a mass of facts into a coherent explanation). This is an embodied-situated process whenever these clue-like characteristics are predominantly material qualities of external signs. Abduction is especially near to the conceptualization of embodiment because it is an inference which relies on a mass of perceptual data for its conclusion.

5.4 The Representational Effect Is Iconic and Abductive

The iconic and abductive role of artifacts in external semiosis can be demonstrated by isomorphic representations of the same problem. These representations embody the same formal structure of problem states and rules, but they present them differently to an agent, allowing for different abductive inferences on how to engage with the problem. In embodying a problem for a situated agent, artifacts iconically define the problem. Different icons of the same problem structure may derive different ways of engaging with the problem, even changing the problem's difficulty. We exemplify this process taking advantage of Zhang and Norman's (1994) experiments with isomorphs of the Tower of Hanoi puzzle.

The Tower of Hanoi puzzle is used in various fields to study problem-solving strategies, recursion, and algorithm design. It's a simple yet challenging puzzle that requires careful planning and logical thinking to solve efficiently. It consists of three poles and multiple disks, each with varying diameters and a central hole for stacking on the poles (see Fig. 5.1). The diameter of these disks establishes a hierarchical order, dictating how they can be arranged and moved between the poles. The primary objective of the game is to transfer all the disks from the first pole to the last while adhering to a set of fundamental rules. In the specific version of the game under consideration here, these rules are as follows: (1) Only one piece can be transferred at a time; (2) A piece can only be transferred to a place in which it will be the largest; (3) Only the largest piece of a place can be transferred to another place. The formal structure of the game is composed of the pieces (disks, for example), places (poles), hierarchy (disk diameters), rules, initial state, and goal state.

Zhang and Norman (1994) employed the Tower of Hanoi puzzle as a means to explore the impact of representations on cognition, specifically focusing on the Representational Effect—a phenomenon where differences in cognitive behavior

Fig. 5.1 The classical version of the Tower of Hanoi puzzle, with three poles and several disks stacked from the largest, in the base, to the smallest, in the top. In the experiments treated here, this order was altered: larger pieces should be put on top of smaller pieces. (Image taken from Wikimedia Commons)

Table 5.1 Rules of the Tower of Hanoi, experiment 2

1—Only one piece can be transferred at a time
2—A piece can only be transferred to a place on which it will be the largest
3—Only the largest piece in a place can be transferred to another place

arise solely due to representational characteristics. To investigate this effect, they conducted a study involving the comparison of performance when using isomorphic representations in problem-solving tasks. These representations contain the same amount of information about the task structure, but differ in how that information is presented.

Zhang and Norman conducted tests examining various levels of isomorphism in representations, including object representations, dimensional representations, rule representations, and problem space structures. The experiment of particular interest to us is experiment 2 (Zhang and Norman 1994, pp. 20–23), which focuses on the level of rule representations. At this level, the rules of the game can be represented in two distinct ways: (i) they are explicitly provided in instructions and memorized by the players, or (ii) they are inherently embedded in the available move options provided by the game materials. These rules, introduced via (i) and (ii), are referred to as internal and external rules, respectively, and are maintained during problem solving either in the players' memory or within the game materials themselves.

In this experiment, three rules governed the game (see Table 5.1), and these rules could be introduced in either an internal or external manner. Three distinct

Table 5.2 Isomorphic representations of the game's formal structure

	"Oranges" (I123)	"Donuts" (I12 E3)	"Coffee" (I1 E23)
Pieces	Balls	Disks	Cups filled with coffee
Places	Plates	Poles	Plates
Hierarchy	Size of balls	Diameter of disks	Sizes of cups
Rules	1—Instruction 2—Instruction 3–Instruction	1—Instruction 2—Instruction 3—Material	1—Instruction 2—Material 3—Material

isomorphic artifacts (see Table 5.2) were employed, each corresponding to a different scenario: "waitresses and oranges", "waitresses and donuts", and "waitresses and coffee". These isomorphs used different materials to represent the rules of the problem. In the "oranges" version, balls ("oranges") served as the pieces, plates as the places, and the ball sizes represented the hierarchy of pieces. The "donuts" version featured disks ("donuts") as pieces, poles as places, and disk diameters indicating the hierarchy. In the "coffee" version, cups filled with coffee were the pieces, plates were the places, and cup sizes represented the hierarchy. Each of the three rules could be either internal (provided as a list of pre-experiment instructions memorized by the players) or external (automatically inherent in the constraints of the game materials). In the "oranges" version, all three rules were internal (I123). In the "donuts" version, rules 1 and 2 were internal, while rule 3 was external (I12 E3). In the "coffee" version, only rule 1 was internal, and rules 2 and 3 were external (I1 E23). The "oranges" version was entirely internal in relation to all rules because the balls on plates could be freely moved without constraints. In contrast, the "donuts" version was external concerning rule 3 because only the top disk could be moved when stacked in poles (unless multiple disks were taken, in which case internal rule 1 was violated). The "coffee" version was external for rules 2 and 3 because stacking smaller coffee-filled cups on top of larger ones would result in coffee spillage, a situation deemed unfavorable in the scenario presented. Consequently, rule 2 was also externalized in that scenario.

The experiment assessed several factors for each of the three isomorphs, including the time taken to solve, the number of steps needed for a solution, and the frequency of incorrect moves. Across all three cases, the most internally represented version (oranges) exhibited the poorest performance, requiring more time, more steps, and yielding more incorrect moves (Zhang and Norman 1994). Conversely, the most externally represented version (coffee) demonstrated the best results, with shorter solution times, fewer steps, and minimal incorrect moves. The donuts version occupied an intermediate position. This experiment, in conjunction with others presented in the same article, led the authors to propose that more externally oriented representations tend to be more efficient for problem-solving (see also Zhang and Norman 1997, 2000).

Zhang & Norman's criteria for distinguishing between internal and external rules align with a measure of iconicity, specifically, the dependence on artifact properties, or S-dependence. Each problem solving process in the experiment can be considered a semiotic process involving the communication of a form or habit from an object to an interpretant through the mediation of a sign. In this triadic relationship, the object (O) is the formal structure of the problem, which remains consistent across all isomorphs. The sign (S) is the artifact through which the game is played. The interpretant (I) is the gameplay, as constrained by the S-O. When we apply this framework and consider the relative dependence of these terms, we can draw the following conclusions for both the internal and external rules:

- Internal rules: The formal structure of the game (O) remains independent of the artifact used for play (S). Altering the artifact does not affect the fundamental rules of the game, which remain consistent. The relationship between S and O cannot be established solely through these two elements; it necessitates the mediation of a third term (I). The alignment of specific play materials with the formal game structure only occurs as a cognitive constraint within the player's behavior during gameplay. Since the S-O relationship depends on I, this exemplifies a case of symbolic semiosis.
- External rules: The game exhibits S-dependence. If you modify the artifact used to play, it directly impacts the formal structure of the game. The relationship between S and O is established independently of the third term (I) because the constraints imposed by the artifact inherently reflect the formal game structure, in advance of the gameplay. Given that the S-O relationship depends on S, this represents an instance of iconic semiosis.

The definitions for this particular case can be generalized to any other case of externalization of constraints. To say that a representation is external in respect to some constraints already implies that these constraints are artifact-dependent (S-dependent), and that we are dealing with iconic semiosis.

To understand the role of abduction in this process, we emphasize the inferential activity required for each move in the game. Solving the game necessitates the player's ability to infer how to progress from an initial state to a goal state while passing through intermediate problem states. In this process, the player continuously makes inferences when deciding how to transition from one problem state to the next. These transitions adhere to the predefined rules of the game, offering the player a range of possibilities to choose from. This inferential process is fallible, in the sense that it doesn't always lead to the optimal gameplay solution. Instead, it has to involve the abductive generation and selection of possible moves based on a set of constraints.

This abductive process is affected by how well one can visualize or imagine the space of possible moves available to him/her at any given moment of the puzzle. A problem space which is more constrained, that is, in which there are less possible transitions between problem states, is also one in which the cost of choice is reduced. When a player is faced with 2 or 3 possible moves clearly indicated by the material at each step of the puzzle, he/she is more likely to choose the optimal move than when faced with 4 or 5 of them. More material constraints that are also perceptually salient result in less options to be considered and play a role in reducing the cost of abducing the next optimal move. In this sense, externalized rules serve as material selectors of possible moves, outsourcing the cognitive costs of performing abductive inferences.

When a player of the cups of coffee version of the puzzle avoids a bad move because of the anticipation of coffee spilling, part of the work of abducing the optimal move is being performed by the material/perceptual qualities of the puzzle. In comparison to the oranges version of the game, which allows several wrong moves to be physically performed, the artifacts in the cups of coffee isomorph transform the inferential activity of figuring out the best move between problem states into a simple perceptual task. Here, the iconic representation of rules facilitates abduction.

5.5 The Representational Effect in Ill-Defined Problems

The Tower of Hanoi is an example of a well-defined problem, characterized by a closed solution set. The puzzle has a predefined goal and problem space, making problem-solving a matter of navigating within this predefined space to discover the regular (rule-allowed) moves that lead to the established goal. However, it's important to note that most everyday problems do not fit this neat structure. In the realm of ill-defined problems, the solution set, the rules, and the problem space may be open and not predefined. Agents tackling such problems are tasked not only with finding a way to reach a predetermined solution, but also with deciding what kind of artifact configurations count as satisfactory solutions. While a well-defined problem solving process already exists as a coherent whole involving rules, initial state and goal state, an ill-defined problem solving process involves the emergence of a structure involving a goal state in relation to rules and initial state. This is a holistic kind of emergence whose product is a regular pattern of organization in which the goal state is justified by rules in relation to the initial state.

The emergent character of ill-defined problems helps to explain why they are often associated with creativity. An often-cited example of an ill-defined problem is the alternate uses test, originally designed by the American psychologist J. P. Guilford to measure a person's divergent thinking abilities. For instance, a participant is presented with a simple object, say a paperclip, and asked to generate as many novel uses for it as possible within a limited time frame. In this case, the paperclip becomes an iconic artifact, whose properties have the power to abductively suggest new uses that redefine the conventionally understood function of the paperclip. For example, if the participant comes up with new uses for the paperclip like a miniature sculpture stand, a cord organizer, or a jewelry clasp, these imaginative solutions amount to a simultaneous redefinition of rules (what one is allowed to do with a paperclip) and solutions (uses for a paperclip) determined by the properties of the artifact (paperclip). In this redefinition, the paperclip is used as an icon that supports abductive inferences. In contrast to well-defined problems with predefined solution sets, in which iconicity is limited to reframing a previously existing problem structure, in an ill-defined problem the iconic and abductive action of artifacts participates in bringing forth an emergent problem structure.

5.6 Iconicity and Abduction: The Semiotic Role of Artifacts in Situated Problem Solving

The representational effect in situated problem solving effectively illustrates Peirce's theoretical proposition that artifacts are Firsts (signs or representamens) of the semiotic process. In the context of situated cognitive processes, a problem always finds instantiation through an artifact in which it becomes capable of producing interpretants. It is this artifact that frames the problem, materializing it in cognition and rendering it actionable. This initial framing shapes the entire process of problem solving: it constructs (or transforms) a problem space, within which actions towards a solution can be abductively inferred. Iconicity plays a pivotal role in this context, as it underlies the embodiment of a problem in the sensorial and material properties of an artifact. Icons and abductions are the kinds of semiotic relations involved in constructing spaces of possibilities for meaningful action.

Framing through iconicity and abduction provides conditions so that the process of problem-solving may unfold as manipulation of the artifact in accordance with the problem's rules. This phase, logically subsequent to framing, centrally involves the imposition of constraints on the artifact, so that changes in artifact configuration maps onto problem states and hence manipulation of the artifact is capable of revealing the solution of the problem. Iconicity once again assumes a central role in this mapping between problem constraints and artifact constraints. Through iconicity such mapping occurs almost automatically because the artifact's constraints are inherently analogous to the problem's constraints that need to be considered. In a case such as the "cups of coffee" isomorph discussed earlier, these constraints are iconic in a manner that greatly facilitates the abductive process of determining the best course of action. As experimental results of Zhang and Norman (1994) illustrate, the iconic representation of the rules in the cups of coffee isomorph almost trivializes the abductive inference of optimal moves by the agent.

Abduction and iconicity correspond respectively to the categories of inference and sign processes in which the situated aspect of Peirce's conception of mind is most directly conceptualized. They model how artifacts, in their materially extended properties, are Firsts of the semiotic relation. Through iconicity and abduction, artifacts construct problem spaces in which actions have a meaningful import towards a solution.

5.6 Iconicity and Abduction: The Semiotic Role of Artifacts in Situated Problem Solving

[illegible]

Chapter 6
The Object of Iconic Artifacts: Why an Underground System Is a Semiotic Problem

6.1 Shaping Urban Minds

It is often mentioned that the icon is a particularly important class of sign when we talk about phenomena related to modeling, invention or discovery in empirical and formal sciences. It is also important when we talk about the arts and literature. We attribute to icons a certain creative power that only this type of sign seems to possess, and which is not directly associated with indexes or symbols. The thesis that significant discovery-related activities rely on the manipulation of sign materiality is receiving increasing empirical support in recent research, as well as new theoretical treatments.[1] When associated with a more strictly semiotic approach, the thesis assumes an explicit semiotic externalism.

The creative power of iconic semiosis is intrinsically linked to the distinctive relation between artifact and object that characterizes an icon. The icon is the kind of sign in which an artifact directly embodies properties which the object also embodies. Iconicity corresponds to cases in which the sign and object are conflated in some respect or capacity. Thus, the iconic artifact has certain determinative power over the semiotic object in its capacity as a semiotic object. Manipulation of the iconic artifact allows characteristics of the object to be uncovered, or, in creative cases, it allows characteristics of the object to be transformed. In this chapter, we explore such a case: how Harry Beck's famous redesign of the map of the London Underground transforms the London Underground itself as a semiotic object. Beck's diagram is often considered as the direct origin of the nowadays dominant

[1] For a review: Paavola (2011) on icon-diagrams and their connection to abductive inference; Becvar, Hollan and Hutchins (2008) on diagrams in chemistry. Andy Clark (2008) devoted a chapter ("Material Symbols") in his book *Supersizing the Mind* to external representation. Many years before, Donald Peterson (1996) had been conducting research on the role of semiotic structures and their relationship with complex cognitive processes.

P. Atã, J. Queiroz, *Mind as an External Sign: A Semiotic Externalist Philosophy of Cognitive Science*, Synthese Library 531,
https://doi.org/10.1007/978-3-032-08988-5_6

representational paradigm of transport systems as schematic networks of lines and stations. Our argument here is that Beck's redesign has reframed the task of navigating the underground transport system. By transforming the Underground system as a problem for its users, the diagram transforms its semiotic objecthood.

The London Underground Diagram is an example of a specific kind of iconic sign, the diagram. Peirce identifies this type of sign as a hypoicon, which, unlike the pure icon, which is just a "possibility", is an instantiated icon that represents existing relations. Diagrams are hypoicons that represent the relationships between the parts of an object through the connections among their own parts (CP 2.277). They can be maps, graphs, lists, tables, musical scores, or any type of formalism used to express relations between their parts. According to Stjernfelt (2022),

> To Peirce, diagrams comprise all structures that are fit to make deductive inferences from. In the topographical map, e.g., you may trace possible routes from one location to another – thus inferring an information which was only implicitly present in the diagram. The routes thus found are truly there as real possibilities in the landscape, given that the map as a premise is, in itself, a true representation of the landscape (if the map is false, of course, you cannot be sure to draw true conclusions from it). Thus, Peirce's argument is that diagrams form the proper generalization of the various formalisms of deductive logic.

According to Peirce, manipulability is crucial for retrieving implicit information from diagrams, whether real or imagined. By manipulating a diagram following certain rules, explicit or not, users can infer new claims that are true, provided that the information put into the diagram is true. This gives the diagram the character of an ideal machine, with the ability to generate new insights and knowledge. Additionally, the possession of implicit information gives diagrams a quasi-mystical quality, evoking the sense of a looming presence of something not explicitly there.

As discussed previously, icons are cognitive artifacts that serve as material tools embedding cognition and shaping our minds. The Underground Diagram has specialized the problem solving of urban commuters as an activity based on extraction and manipulation of relations, capable of generating overall changes in the behavior of the users and influencing the understanding of the city itself. For Peirce (MS [R] 339:286r), "The purpose of a Diagram is to represent certain relations in such a form that it can be transformed into another form representing other relations involved in those first represented and this transformed icon can be interpreted in a symbolic statement. [...] The Diagram is an Interpretant of a Symbol in which the signification of the Symbol becomes a part of the object of the Icon." In our argument, the subway diagram is an interpretant of an emerging semiotic niche of relationships between users and the transportation system. This emerging niche of relationships becomes integrated into the diagram's object.

6.2 The Underground Diagram's Design

Urban transportation maps (underground, buses, trains, etc) all over the world adhere to a similar representational paradigm of a schematic diagram that sacrifices geographical accuracy in favor of a clear picture of the interconnections of lines and stations in a network. One of the first examples of this kind of diagram, and often considered its origin, is the London Underground Diagram (LUD), or tube map, a "form of representation judged to be so effective that it is now employed by virtually every transportation authority in the world" (Spence 2007, p. 77). The original version of the LUD was created by Henry C. (Harry) Beck in 1933. Previously to the LUD, maps of the London Underground System adhered to more literal, or geographically accurate, representations of the lines and station locations. These maps represented the city territory in geographical scale, and the transport lines waving through it. The length, direction and curves of the transport lines were represented coherently, and sought to preserve proportional geographical lengths, distances and locations.

Beck was an engineer draughtsman, and his redesign was inspired from electric circuit diagrams. An electric circuit diagram typically uses a solid line to depict the conductor or wires that establish the circuit's route. Along this route, symbols are placed to indicate the different components of the circuit, such as power sources and resistors. Circuit diagrams serve various purposes, including the design, construction, and maintenance of electrical circuits, as well as the calculation of current in different circuit segments. A pictorial circuit diagram presents the circuit components using images, while a schematic diagram utilizes standard symbols. Both types illustrate the connections between devices, including power and signal connections. However, the arrangement of the components and their interconnections on the diagram may not reflect their physical locations in the final device. Beck's design for his first sketch for the London Underground Map, produced in 1931, was based upon and adapted from electrical circuit diagrams, which

> omit or falsify the relative physical position of wires in order to convey the information about connectivity. Beck saw a similarity with the underground railway network in that it was possible to ignore the geographical information altogether and remove some of the sources of confusion in the previous more literal maps. (Whitby 1996, p. 70)

Beck's innovation was to get rid of proportional scale so as to emphasize connectivity. His design follows an octagonal grid, where color-coded straight lines meet at angles of 90° or 45°. Stations are regularly arranged along these lines, emphasizing the sequentiality of the stations along the line rather than the geographical distance between the stations. The River Thames is also represented by straight line segments following the octagonal grid, and was introduced in the diagram in order to better situate the users in relation to a schematic understanding of the city.

Instead of a geographical scale, the diagram adapts the scale to the number of stations represented (see Garland 1994). The central area of the network and the outer areas do not follow the same proportion. The outward portions, with less stations, were compressed, while the central area, with more stations and lines, appears to be viewed through a convex lens so as to enlarge its scale. The stations are represented differently whether they are interchange or non-interchange stations. In Beck's 1933 diagram, the non-interchange stations were represented as ticks along a line, while the interchange stations were represented as diamonds side-by-side on the meeting of two lines. While the LUD has undergone successive redesigns over the years, the essential innovations of Beck's original diagram were retained.

6.3 The LUD as an Iconic Artifact

The notion of iconicity can be understood in different ways. Traditionally, it has been defined as "similarity" between sign and object. It has also been defined as relative dependence on S in the S-O relation (see Queiroz 2012). Stjernfelt (2011) identifies two different contrasting conceptions of icon and iconicity in Peirce's work: first, the icon can be operationally defined as any sign whose manipulation is able to reveal more information about its object. This operational definition of the icon focuses solely on the capability of a sign to enclose information about its object. Stjernfelt uses the term "operational iconicity" to refer to the conception of iconicity arising solely from this operational definition. Operational iconicity contrasts with a stricter notion of iconicity, in which the sign-object relation has to be considered as including factors such as immediacy of the information presented and economy of elements. We refer to the conception arising from these stricter criteria as "optimal iconicity" (Stjernfelt 2011, p. 400).

Stjernfelt (2011, p. 414) exemplifies the distinction between operational and optimal iconicity through the example of a digital picture. A picture can be digitally represented as pixels on a screen or as a linear sequence of digital information. If we only take into account the operational definition of the icon, the two representations are equally iconic (operational iconicity): they are informationally equivalent (i.e., enclose the same amount of information), and one can be algorithmically transformed into the other. However, this operational definition alone ignores some representational features which are decisive for the S-O relationship in each sign: in the pictorial image, for example, object contours are represented as continuous lines while in the linear digital representation this information is scattered throughout the code. A single object contour is more similar to a single continuous line than several scattered pieces of information (see Stjernfelt 2011, p. 414). A one-to-one correspondence holds some kind of logical and phenomenological intrinsic iconic value that is shattered by a one-to-several correspondence. Therefore, it is more iconic (optimal iconicity).

In the LUD, the operational iconicity criterion tells us that we are dealing with an iconic artifact because a user manipulating the LUD is able to discover implicit information about the underground system, e.g., in which line embark to get to a specific station. However, if the criterion for iconicity is the capacity to enclose information about an object, then we should consider Beck's diagram as less iconic than the geographically more accurate maps that it replaced. A map encloses information that Beck's diagram does not: geographic location of stations, the distance between stations, the real length of lines and the specific directions and changes of directions of maps (see Table 6.1). There is more information to be discovered about the underground system in a geographically accurate map than in the LUD.

On the other hand, the optimal iconicity criterion is able to stress the LUD's specialization as a problem-solving tool, thus differentiating it from other representations equally capable of revealing information about lines and stations, but not equally capable of framing problems about navigation in a certain way. An underground diagram is a problem-solving tool for the problem of finding one's way in underground transportation systems. This problem has a general structure of rules: the initial state is the station from where the user departs; the goal is to know how to arrive at the desired station; the rules to go from station to station are that the trains follow predetermined paths (lines), and that one can change between lines in interchange stations (see Table 6.2). The problem is solved every day by users of underground systems. Solving the problem according to the rules is a habit, a regular pattern of action and meaning, and the underground diagram embodies this set of rules, making them available to users. As with the Tower of Hanoi example discussed in Chap. 5, the LUD is iconic (optimal iconicity) because it embodies the structure of the problem-solving task of navigating the underground system. The geographically accurate map, while it encloses more information about the underground system, does not make this problem-solving structure as directly manipulable as the diagram. Using the LUD makes it easier to locate the user's current location (initial state) and destination (final state), to make hypotheses on how to better reach the destination and to verify these hypotheses.

The iconic character of the LUD is at odds with a criterion of iconicity based on enclosing all the available information about an object. On the contrary, its iconicity

Table 6.1 Comparison between information of O (the underground system) contained in S (maps and diagrams) for the LUD and a geographically accurate map of the underground system. The LUD contains less information about the underground system than the map. Therefore, it is less iconic for operational iconicity, suggesting it to be less similar to the Object. However, it is more efficient, therefore more iconic for optimal iconicity, suggesting it to be more similar to the Object

Information of O accurately contained in S	Geographically accurate Map	London Underground Diagram
Stations	Yes	Yes
Connections between stations (tube lines)	Yes	Yes
Connections between lines (interchange stations)	Yes	Yes
Distance between stations	Yes	No
Geographic location of stations	Yes	No
Length of lines	Yes	No
Specific directions and changes of directions of lines	Yes	No

Table 6.2 The formal structure of states and rules of the problem solving task of navigating the underground system

Problem-solving task: navigating the underground system	
Initial State	the user's current station
Final state	the user's goal station
Intermediate states	every station the user is going to access in order to go from the initial to the final state
Rules for moving between states	In order to move the user embarks on a train, following its path on the line until the station (final or intermediate) she wants to disembark
	The train will follow its path on one particular orientation until the end of the line. It will not change its trajectory, orientation or line while traveling
	There are two types of stations: normal stations only allow for embarking or disembarking on one line. Interchange stations allow for changing lines

lies in removing information superfluous for the problem-solving task, and which would needlessly increase the cognitive cost of problem-solving and make it less manipulable. In this sense, the LUD is less iconic than a map when we consider operational iconicity, but more iconic when we consider optimal iconicity.

6.4 The Object of the LUD

Is the LUD more or less similar to the underground system than a geographically more accurate map? Viewing similarity through the lens of operational iconicity suggests it is less similar. The LUD may be seen as distorting the real underground system, representing straight lines that, in reality, are not truly straight. Conversely, viewing similarity through the lens of optimal iconicity, and considering factors such as manipulability and problem-solving efficiency, suggests that the LUD is more akin to the reality of the underground system compared to the map. For instance, the map, offering a bird's eye view of the city, abstracts away from the actual experience of the underground system for regular users. For someone who relies on the underground daily, city navigation is less about knowing street connections and geographical directions and more about knowing which station to embark and to disembark, and which line changes are necessary to go from one to another. The lived experience of riding the underground car promotes a focus on the diagrammatic over the topographical. When using underground transportation, visual cues for navigating the surroundings are absent, and due to the speed of the trains and the frequent stops, the temporal experience of distance may depend less on the actual line length and more on the number of stops a line contains.

The conclusion above seems to introduce a contradiction regarding the status of the underground system as a semiotic object. In comparison to a geographically accurate map, the LUD is concurrently less iconic for operational iconicity, yet more iconic for optimal iconicity. As these notions of iconicity are non-trivializations of the concept of similarity (see Stjernfelt 2011, p. 397), we can also deduce that the LUD is at the same time more similar and less similar to the underground system than a geographically accurate map.

This apparent contradiction underscores the need to conceptualize the semiotic object as a process, subject to change and evolution, and partly shaped by iconic processes. Peirce's semiotics captures this processual nature through the notions of immediate and dynamical objects, that is, the object as a sign represents it, and the object as a real process which surpasses any single immediate representation by a sign:

> We must distinguish between the Immediate Object – i.e., the Object as represented in the Sign – and [...] the Dynamical Object, which, from the nature of things, the Sign cannot express, which it can only indicate and leave the interpreter to find out by collateral experience. (CP 8.314)

As Queiroz and colleagues summarize: "The dynamical object is something in reality that determines the sign, but can be represented by the sign only in some of its aspects" (Queiroz et al. 2011, p. 101). These aspects are the immediate object, but the dynamical object is a whole that cannot be completely defined by the immediately represented aspects. Establishing the dynamical object from the immediate object is a process that requires the agent's competence:

> Because the sign represents the dynamical object in some of its features only, never in its totality, it can simply indicate that object, and it is left to an interpreter to establish what is the dynamical object through the interpreter's competence as a user of that sign, which, in turn, results from its previous experience and learning to become an interpreter. (Queiroz et al. 2011, p. 101)

We approach this competence in the context of situated problem-solving activity that depends on external iconic signs. Distinct immediate objects stemming from the same dynamical object present distinct interpretive challenges for agents, along with distinct active external drivers for the cognitive processes needed to address those challenges. In this context, a semiotic object closely aligns with the concept of a problem in a problem-solving task. The portrayal of an immediate object through an icon corresponds to the formulation and framing of a problem.

The LUD and the map stand for distinct immediate objects of the same dynamical object, thereby framing the underground system as different problems and providing different means for solving them. The diagram reframes the wayfinding problem in a topological rather than a topographic problem space. While a map implies that the rules for solving the navigation puzzle hinge on the topography of the underground system and the city, the diagram, in contrast, omits and distorts information about distance, scale, and territorial displacement. Instead, it concentrates solely on the diagrammatic pattern of connections between stations and lines. This effectively reduces the amount of implicit information in the diagram, minimizing the range of operations to be performed (such as determining real distances). Beck has introduced features that don't increase the amount of information but rather externalize the inferential processes needed to access relevant information for the task at hand. This means that user behavior and the task itself are constrained and shaped by the material iconic features of the representation. Although this framing of the problem space makes the diagram less suitable for tasks like estimating distances and geographical locality, they prove highly efficient for the specific task of selecting the appropriate connections to reach a particular station. The efficiency gained in the shift from geographically accurate maps to the LUD aligns with how iconicity in the LUD puts users in direct contact with the most crucial rules involved in the wayfinding problem-solving task.

The LUD represents the London underground system faithfully, in its character as a network of connections. In this context, the diagram, when contrasted with a topographically more precise map, is not in any sense a distorted portrayal of the city, nor is it less faithful to the reality of the underground system. The city, as an environment, encompasses a variety of lived experiences and qualities, and the diagram and the map embody different possibilities of how the transportation system, and the city as a whole, may shape habitual effects for a population of agents. The diagram reinforces and provides access to a certain habit of the city:

> [The diagram] represented a view of London and its Underground that people had – albeit unconsciously – been looking out for. One that would cope with the information needs of a growing population now committed to travelling considerable distances to work every day from the new suburbs, and of visitors from the provinces making their first, timorous acquaintance with the metropolis in increasing numbers throughout the 1930s. (Garland 1994, p. 7)

Beck's design has diminished the resemblance of the LUD to some topographic characteristics of the underground system, opting instead to enhance its similarity to a specific structure of rules and goals that underlie a certain experience of urban transportation and urban space. It has chosen a habit—a set of relations and rules of action, embodied in an iconic artifact, so that it reappears, as iconic semiosis, in the behavior of the users. This habit thus becomes an integrated part of the minds of Londoners, now hybrid beings embedded with a particular set of rules of action. For them, the LUD stands as a commonly familiar model, a specialized environment built for the extraction and manipulation of relations. In this sense, the impact of the efficiency of the LUD goes beyond the scope of discrete problem-solving tasks. It becomes part of the cognitive/semiotic niche of urban dwellers, making them more attuned to the urban environment and influencing their overall behavior and perception towards the city. It is "more than a simplification of Underground railway routes (…) it is an essential simplification of the city itself" (Garland 1994, p. 5).

6.5 The London Underground Diagram as a Semiotic Object and Problem Space

The London Underground System is a problem. The generalization of this thesis has strong ontological implications. Semiotic objecthood is indissociable from posing problems. For anything to be a semiotic object is to manifest to an interpreter in the form of an interpretative challenge that constrains the action of semiotic artifacts. There is a two-way dependence between artifacts (signs) and problems (objects): on the one hand, objects can only manifest through signs as problems that constrain artifacts; on the other hand, artifacts are only really artifacts if they stand in a triadic connection to problems they embody and to the solutions whose conditions for emergence they provide. If this thesis is correct, it suggests a processual ontology in which things, substances, structures or properties can be more accurately characterized as problems.

If an environment is a certain way to put together the bits and pieces of a physical world into a lived experience, the underground diagram participates in putting together the bits and pieces of modern cities into lived environments. However, the diagram also influences the physical space by modifying the behavior of the users (who are themselves an important physical element that constitutes the underground system), endowing them with a system of rules to navigate around.

In our approach, while it may be of little relevance whether cognition is happening inside or outside the head, it is decisive that it must happen in artifacts that stand triadically for problems and solutions. This conception neither restricts representations to symbolic semiosis as in orthodox representationalism, nor rejects representations as in anti-representationalism. The study of distributed cognition benefits from the system proposed by Peirce in the sense that it offers a model of how and by virtue of what the mind semiotically unfolds itself. As the study of the representations and its functioning becomes a necessary part of the study of cognition, Peirce's conception of icon arises as an important tool for the investigation of thought processes.

Iconicity is a central idea that connects cognitive distribution, niche construction and representational efficiency. An efficient representation is an icon of a structure of habits (rules of action) that foster certain kinds of cognitive behavior that are appropriate for an objective (here conceived as a game-like activity with an initial state and a goal state). Iconicity helps to clarify how it is possible for a habit to be embedded on a representation and to constrain the interpreter. Representational features act themselves as rules of action because of the interrelatedness of its parts being analogous to certain effects of the environment which allow: (i) the embedment of extractable information in the sign about the object (related to operational iconicity) and (ii) the direct manipulation of this information (related to optimal iconicity). Through iconicity cognition is distributed. As representations mold cognitive behavior, they become part of an ongoing process of niche construction, where the cognitive potentialities of groups of individuals are expanded or directed towards certain purposes. In our example, a particular experience of urban transportation, partly determined by the technology itself of Underground transportation, materializes itself on a sign that causes urban dwellers to adapt to it, thus participating in niche construction. The most decisive step of the process happens through iconic semiosis. The reduction of the amount of information in a representation by virtue of its specialization for specific tasks does not oppose different conceptions about iconicity, but rather redefines the object of the sign, clarifying its role as the materialization of a problem space optimized to function as an environment where cognition develops through manipulation of diagrams.

Chapter 7
Poetry Improvisation as Problem-Solving in Distributed Cognitive Systems

We propose a semiotic externalist approach that takes cognition as semiosis, gives central importance to the notion of temporal distribution, describes the elements of distributed cognitive systems (DCSs) as signs, and identifies the DCS itself as a system that enacts a sign. This is a semiosis-centered, and thus a non-agent-centered account of DCSs. In order to develop and illustrate our argument, we describe an example of DCS—the Brazilian verbo-musical improvisational tradition of *repente*—considering it first as embodiment of the formal structure of a cognitive task, and then as embodiment of a semiotic process. The latter corresponds to a semiotization of the description of *repente* sessions as DCSs, that focuses on how the DCS can embody a metasemiotic process, semiosis that supervenes on, and determines, distinct types of smaller-scale semiotic process.

7.1 Introduction

In the previous chapters we have explored the role of artifacts and problems within a view of semiosis as problem solving. In this chapter we explore who or what is the cognitive system solving problems. Our starting point is the notion of distributed cognitive system (DCS), and we investigate the relation between DCSs and the action of signs.

Distributed cognition is an active externalist approach that shifts the focus of cognitive science from the heads of individuals to *distributed cognitive systems* (DCSs) of agents, situation and artifacts (see Hutchins 1997; Sutton 2010; Davies and Michaelian 2016). Reacting against the internalist neural-computationalist paradigm that detaches cognition from spatio-temporal situatedness, distributed cognition is concerned with cognitive processes as they happen "in the wild" (Hutchins 1995a). Instead of describing cognition as it happens in the brains of individuals,

P. Atã, J. Queiroz, *Mind as an External Sign: A Semiotic Externalist Philosophy of Cognitive Science*, Synthese Library 531,
https://doi.org/10.1007/978-3-032-08988-5_7

distributed cognition tries to describe the operation of DCSs, broader than individual brains. Thus, for example, the cognitive operations involved in flying a commercial airliner can be described in terms of a DCS that includes the crew (pilot and co-pilot), the flight instruments, and other external media of representation (such as plastic cards), all used according to conventionalized procedures (Hutchins 1995b). A biomedical research laboratory can be described as a DCS, composed of several researchers, various different pieces of equipment, books and online reference material, etc. (Nersessian et al. 2003). An Elizabethan theatre company can be described as a DCS whose workspace is the theatre stage itself, and includes components such as relevantly-positioned stage-doors, large sheets of paper displaying the structure of entrances and cues of a play, as well as the use of versified language (which help memorization) (Tribble 2005). In these examples, some cognitive processes essential for how the system operates happen outside of the heads of individuals, and involve more than one individual. Cognition is external, and crucially depends on cognitive artifacts that aid, support, enhance or improve cognition (Hutchins 1999; Menary 2010).

But what defines and delimits a DCS? This question of definition brings with it a question of ontology, that aims at the real thing(s) or process(es) that grant(s) DCS its criterion of parthood in reality. Is a DCS less real than the agents and artifacts that compose it? If one defines a DCS as a composition of agents and artifacts, that's what the definition suggests. Or is a DCS real, in itself, in a different way than the reality of the agents and artifacts that compose it? If so, how to define this real entity or process?

Candidates to answer this question include notions such as cognitive agent, cognitive tasks, task-relevant information flow, and information processing. We suggest here another alternative: a *semiosis-centered* approach to DCS. In our approach, processes such as semioses are ontologically real, that is, semiosis is not just a name to describe what other real entities do. Semiosis should be conceived as a real process, operative in nature in widely different scales, such as DNA transcription, human and non-human cognition, cultural and biological evolution (Hoffmeyer 1997). To say it is a real process means that semiosis is not just a name to categorize together similar mechanisms observed in different processes. Rather, these different processes are all instantiations, in widely different scales and environments, of the same fundamental process.

If semiosis is a process, it is neither a thing, an agent, nor is it the action performed by a thing or by an agent. In the same way, the term semiosis is not a way of reworking, reconceptualizing, or rebranding notions such as information processing, computation, or interaction among agents and artifacts. Semiosis emerges from the interaction of agents and artifacts, but is not identical nor reducible to them. We should recall two properties of semiosis that will be especially important for our argument in this chapter.

The first property is triadicity, or triadic irreducibility. Semiotic processes are composed of tri-relative terms. Differently from dyadic, or reactive, events, triadic actions have the power to beget more of themselves, in the form of a growing general pattern, or disposition to act, called habit. The tri-relative action of semiosis involves a first relational term (sign, or artifact), whose action is not free, but constrained and determined by something external to it (object, or problem), but only insofar as this sign-object relation has the capacity to re-instantiate itself as an effect in a broader situated system (the system, here, is called an interpreter—a relational term which should be understood much broader than a person or conscious mind[1] —, and the effect is called the interpretant). In problem-solving, a solution for a problem is an interpretant of the artifact-problem relation. In an irreducible triad, what brings together all the terms of the relational complex cannot be any sum of dyadic relations between the terms. Any relation between a sign and its object depends on an interpretant, or, any relation between an artifact and the problem it embodies depends on how the artifact is used in problem solving, whether potentially (immediate interpretant), actually (dynamic interpretant) or as a regulatory tendency (final interpretant). This triadic irreducibility implies some level of agential autonomy: semiotic processes are composed of actions that self-organize towards self-reproduction, in the form of habits.[2] Thus, signs are not defined as being tools employed by individuals to realize their own (individual's) internally motivated goals and purposes, but, rather, that interpreters (which can be conceived as distributed processes, whether chemical, biological, geological, and so on) embody, at any time, several ongoing processes of self-reproducing habits.

The other property is dynamicity. Signs are not "things". The mode of being of signs is to act—determine effects—in the world. Stability of semiosis in time is not understood as stasis (such as in static representational states) but as regularity of sign action across an always ongoing communication. Semiosis is a dynamic process, with signs continuously translating into interpretants which in turn act as signs to produce yet other interpretants. As Savan (1986, p. 134) argues, an interpretant is both the third term of a given triadic relation and the first term (sign) of a subsequent triadic relation. This is the reason why semiosis cannot be defined as an isolated triad; it necessarily involves chains of triads, which throughout its development continuously grow a self-organized and self-replicating agency.

Active semiotic externalism involves a kind of cognitive distribution. Peirce's cognitive semiotics is concerned both with spatial and with temporal distribution of cognition, but the most central and defining properties of semiosis concern temporal distribution: "To say, therefore, that thought cannot happen in an instant, but requires a time, is but another way of saying that every thought must be interpreted in another, or that all thought is in signs" (CP 5.253). Semiosis is a time-dependent process: an action of signs which develops in several phenomenological, historical and evolutionary scales. To describe a "sign" is to describe a process of relation whose properties are irreducible to any collection of synchronic moments or components. Semiosis is embodied in biotechnical systems as a tendency to generate

[1] The interpreter is described by Peirce as a "Quasi-mind" (CP 4.536), a description which demands, for its proper interpretation, a clear recognition of Peirce's broad concept of "mind" (Ransdell 1977).

[2] For a recent treatment of the phenomenon of "semiotic agency", see Sharov and Tønnessen (2021).

problems, artifacts and solutions from a past history of interactions, a process that involves both anticipation of the future and generalization of the past history.

In our usage, Peirce's cognitive semiotic cannot be dissociated from a philosophy (epistemology and ontology) of processes. Signs, objects, and interpretants are not things made out of property-bearing substances, but parts of a general process that develops throughout many occurrences of changes. Semiosis is always in process. This processualist understanding is precisely what allows us to dispense with the notion of agent as the ontological center of a distributed cognitive system. We don't start with the premise of a static system, so it is not necessary to postulate an agent who is the causal originator of changes and transformations in the system. In other words, signs don't need to be put into action by the will or choice of an agent. As processes, the mode of being of signs is to be in action, and sign action already presupposes some level of autonomous agency of signs.

In order to present a semiosis-centered approach to DCSs, we juxtapose and contrast two descriptions of the same phenomenon of distributed cognition—the verbo-musical improvisation of the Brazilian *repente* tradition. *Repente* takes the form of a challenge in versified sung poetry. It can be described as a multiagent, distributed cognitive system that involves the poet-singers who improvise, the audience, and several cognitive artifacts. We describe *repente* sessions first as task-relevant information flux, or DCSs that embody the formal structure of a cognitive task (following Davies and Michaelian 2016),[3] and then as semiotic processes. The latter corresponds to a semiosis-centered description of *repente* sessions as DCSs.

7.2 Repente

Repente[4] ("sudden", in Portuguese) is a Brazilian tradition of verbo-musical improvisation that takes the form of a challenge in versified sung poetry. This kind of popular improvisational challenge is an ubiquitous phenomenon, manifesting itself under various forms across the world. Câmara Cascudo (2004), a pioneer in ethnographic and anthropological studies of the *repente* in Brazil, describes the *repentista* (the practitioner of *repente*) as a descendent the Greek aedo, the Anglo-saxon greeman, the Arabian moganis and meris, the armorican bards, the scandinavian skalds, the minstrels, and the troubadours of the Middle Ages. *Repente* exists in the northeastern region of Brazil, at least since the mid nineteenth century, having emerged in the region of Pajeú, in the state of Pernambuco and Cariri, in the state of Paraíba, in cities such as Itapetim, São José do Egito, and Tabira, in Pernambuco, and Monteiro, and Patos, in Paraíba, among others. These cities form what is known as the "quadrilateral of sung poetry".

---

[3] For a detailed discussion of this article and a highly technical treatment of "cognitive bloat," see: Gillett (2017).

[4] Internet video searches for the terms *cantoria* and *repente* should provide easily accessible examples of *repente*.

Practitioners of *repente* are called *repentistas* (because they improvise), *cantadores* ("singers", in the sense that any topic can be sung by them) or *violeiros* ("guitar-players", *viola* being the name of their typical guitar). These different names point to different skills and activities of a *repentista*: improvisation, singing, and guitar playing. The names *repentista* and *cantador*, in particular, stress different aspects of the improvisational activity:

> One is considered to be a "great *cantador*" if he/she follows a sentence structure [uninterruptedly] when planning the verses, is knowledgeable and displays this general knowledge skillfully in relation to the embedded rules and models of poetry. Differently, the "great *repentista*" is the one notable for fast reactions to what the partner sings and to what is happening in the immediate environment of the event, often starting his/her verses without any certain script — the focus here is on the situational aspect of the poetic practice. Every *cantador repentista* (in the general sense) is *cantador* and *repentista* (in the strict sense), but there are those who stand out more in one or other dimension of this art. (Sautchuk 2009, p. 69, translated from original in Portuguese)

In a *repente* performance there are most often two poet-singers each equipped with his/her own guitar. They take turns generating verses according to the rules. The metric constraints in *repente* are embedded in the distinction between genres. *Repente* has a number of different genres. The largest count is of around 50, although circa of eight or ten genres are the main ones, more widespread and more commonly performed (Oliveira 1999, pp. 52–53). What a genre informs is, mainly, a metric structure of syllables per verse line, distribution of accents inside a verse line, number of lines in a stanza and structure of rhymes in a stanza. *Repente* is a competitive challenge between the *repentistas*. The singers must display their skill by improvising metrified and rhymed verses, according to the theme developed in the performance, and in dialogue with each other (Sautchuk 2011). The audience plays a major role in the evaluation of the performers (Sautchuk 2009), reacting negatively to mistakes—such as hesitations, metric inconsistencies, overused rhymes, thematic repetitions, difficulty in developing a theme or in responding to the opponent —, and highlighting skillful moves—such as creative thematic developments and twists, perfectly metrified verses delivered fluently and without hesitation, quick reaction time, and provocative personal challenges and answers to these challenges. *Repente* sessions are called *cantorias* (which roughly means "singing sessions"). Organized festivals of *cantoria* are a common occurrence. In these festivals, the competition is formalized, with themes randomly assigned for the competitors, and a specialized jury that proclaims a winner (Sautchuk 2009, pp. 133–136).

7.3 *Repente* as a Distributed Cognitive System

A *cantoria* can be described as the operation of a distributed cognitive system. Such a description highlights how a *cantoria*, as a cognitive and cultural practice, is situated in an environment of cognitive artifacts. Cognitive externalism considers cognitive artifacts as active non-biological components of cognitive processes, with agents and artifacts integrating DCSs. But there are different views on the ontology of DCSs (Davies and Michaelian 2016). An agent-based or agent-centered view considers DCSs as extensions of an original center, which is an individual cognitive agent. This is the case of Andy Clark and David Chalmers' extended cognition (1998), according to which cognitive agents augment and extend their minds by coupling themselves to external artifacts. The focus, in this case, is on a system centered on a single agent (Clark 2008, Chap. 6). For Hutchins (2011), Clark's agent-centered externalism fails to account for the crucial role of culture in cognitive externalism. An enculturated account of distributed cognition focuses not on how cognition extends from an agent to artifacts around him/her, but on how multiple agents and artifacts are organized in the context of cultural cognitive ecosystems (Hutchins 2011, 2014). In line with this view, Hutchins (2014, p. 36) characterizes distributed cognition as an emergentist approach to cognition.

In contrast to an agent-centered approach to DCSs, Davies and Michaelian (2016) propose a task-based approach. A task-based approach for individuating cognitive systems starts by identifying a cognitive task (which Davies and Michaelian characterize as information processing tasks) and then includes, as part of the same DCS, whichever agents or artifacts participate in task-relevant information flow for the purpose of solving that task. In Davies and Michaelian's description, if information flows between entities, and the information is actually used to solve a cognitive task, then these entities—regardless of being internal or external to the bodies of agents—can be considered parts of the same DCS in regard to that task. Davies and Michaelian point out that this task-based approach for individuating DCSs allows scaling up and down the level of observation. Tasks can be broken down into subtasks, or seen as parts of overarching tasks with overarching goals. As, in this approach, tasks define DCSs, scaling up and down tasks allows scaling up and down the respective DSCs that enact them. A DCS can be scaled down to focus only on some of the neural structures in the brain of an individual human agent, or scaled up to include multiple individual human agents, interacting to solve overarching cognitive tasks. Thus, the same individual human agent can participate in different DCSs, and a same DCS can include several individual human agents.

In the case of *repente*, and following Davies and Michaelian (2016) task-based approach to distributed cognition, a DCS can be identified on the basis of the cognitive task of performing the improvisational dialogue of a *cantoria*. Subtasks include, for example, composing (on the spot), singing, playing the guitar, and evaluating sung verses. While the overall task of the improvisational dialogue defines the *cantoria* as a whole as a DCS, these subtasks define smaller-scale DCSs, such as an individual singer with his/her own guitar, the pair of singers, and the audience. We are interested here in the scale of the *cantoria* as a whole.

The cognitive task of a *cantoria* can be described as a problem-solving task. In the classical approach to problem-solving, problems can be seen as determining, for their solvers, an abstract formal structure of the problem that is invariant across different instantiations of the task. A different approach is situated problem-solving, which is highly concerned with local and contextual influences to tasks, so that the formal structure of a problem is activity-specific and materially-dependent, and varies according to the artifacts available to the solver at a given situation (Kirsh 2009, p. 271). *Cantoria* is an example of an ill-defined and situated problem-solving task. It is ill-defined because it is a creative task without a closed set of solutions, although it possesses a fixed set of rules. It is situated because the formal structure of the task is embedded in the artifacts used: the poet-singers frame and solve problems with and through their poetry and guitar-playing. There are at least four kinds of constraints that constitute the formal structure of the cognitive task of *cantoria*, exemplified below:

- Temporal constraints: *repentistas* cannot take too long to compose their verses. When one *repentista* finishes singing a stanza, the other *repentista* has a few seconds to start singing his/her own, otherwise he/she is perceived as hesitating.
- Metric and rhyme constraints: *repentistas* have to respect a predetermined number of syllables in a verse, positions of accents within a verse, the number of verses in a stanza, and the structure of rhymes in a stanza.
- Thematic constraints: a duo of *repentistas* has to dialogue around a theme, and has to develop this theme throughout the performance. A *repentista* has to have something to say about the theme, adding new facts or twists to it in a creative way. This thematic constraint is also a way to prevent the singing of memorized verses. Because the *repentistas* don't know in advance what themes they are going to sing about, they cannot prepare and memorize specific verses about it.
- Dialogical/situational constraints: the *repentistas* have to dialogue, to address each other and react to each other's verses. Answering one another and reacting to an immediate situation is a display of reasoning speed and improvisational skill.

These constraints define the *cantoria* as a problem and as a semiotic object, and are iconically embodied by cognitive artifacts which facilitate abductive inferences, in a process of cognitive offloading. The *repentistas* use specific tools to offload these problem-solving constraints. Such tools are emblematic of the *cantoria* practice: the *viola* (guitar), the *toadas* (specific melodic and prosodic structures), and the oral poetry tools of the *motes* ("mottos"), and the *deixas* ("cues", "hooks" or "catchwords").

The *viola* is the characteristic guitar of the *repentistas*, and is responsible for the instrumental musical accompaniment of the sung verses (Sautchuk 2009, pp. 37–39). In a *cantoria*, the *viola* is used to play a *baião-de-viola*, a repeating cycle of two alternating chords (typically A and E), with small variations. Eventually, short melodic fragments can be played, but they are not developed further. Most often, the *repentista* plays the *viola* while the opponent is singing, as well as during the short interval when no one is singing, but stops playing when it's his/her own time to sing. In this case, the period when the *repentista* plays the *viola* is precisely the period he/she is composing the conclusion of the next stanza (the *repentistas* typically compose their verses backwardly, rapidly preparing the concluding verses while the opponent sings, and then improvising the starting verses on-the-spot [Sautchuk 2009, p. 85]). The role of the *viola* is seen as crucial for the *repente* activity. Oliveira (1999, p. 98) suggests the *viola* helps the *repentista* to focus his/her mind: "[W]ithout the *viola*, the cantador won't be able to perform: the *viola* is like a weapon, in such a way that modifies the thought of the *cantador*, focusing his/her mind on the verses, the audience, and the environment as a whole."

The *baião-de-viola* helps to offload the metric and the temporal constraints of the performance. It provides the tempo of the improvisation (Oliveira 1999, p. 129) and gives the rhythmic reference for the dialogue, with the beats of the chord cycles aligning with the metric beats of the verses (Sautchuk 2011, p. 39). The *baião-de-viola* also informs the *repentista* when to start singing. It is common for the *repentista* to start singing after two or three chord cycles. Thus the chord cycles embed the time interval available to the *repentista* for composing the verses, which allows the audience to notice and evaluate how fast a *repentista* is. Furthermore, the *baião-de-viola* performs the phatic function of preserving the rhythm and communication intensity of the performance during pauses or intervals, such as breathing pauses or when a *repentista* needs to stop and think (Sautchuk 2011, p. 39).

The *toadas* are the predetermined melodic lines into which the *repentistas* fit the improvised verses (Ramalho 2001, p. 27). The *toadas* are the main tool used for offloading the metric and rhyme constraints of the improvisation. Sautchuk (2011, p. 79) notes that the *repentistas* learn to metrify verses by learning to hear and imitate sung prosodic and metrical patterns. *Toadas* are importantly connected to the different "genres" of *repente*. Each genre has appropriate *toadas* which conform to the metric demands of that genre. The melodic and prosodic variations of a *toada* constrain the distribution of syllables, accents and rhymes in the desired way. Thus, by finding verses that fit in the melodies, the *repentistas* can respect the metric constraints of a genre, and by varying melodies, the *repentistas* can improvise in different genres. The *toada* externalizes and offloads the metric constraints of *repente*: it transforms calculations about syllables and accents into a simpler task of fitting words into melodic and prosodic slots. The overall structure of the *cantoria* consists in an alternation between the *toadas*, which are sung, and the *baião-de-violas*, which are played on the guitars.

Some genres of *repente* make use of *motes* ("mottos"). A *mote* is a sentence, typically composed of two verse lines, that the *repentista* must use to conclude stanzas (Sautchuk 2011, pp. 28–29). A *mote* is assigned to a duo of *repentistas*, either by audience request or, in the festivals, by some randomization procedure. The mote provides a further constraint, thematic and metric, to the improvisation. The *repentistas* must improvise stanzas that lead, thematically, to the *mote* in the end. In this sense, the *mote* is a device that ensures that *repentistas* stay on topic and explore different aspects of a theme. It offloads the cost of thematic development, encapsulating it in a single problem-solving goal: coming up with a stanza into which the *mote* can fit. The *mote* also establishes metric constraints, since it provides a rhythmic structure that must be followed by the *repentistas*.

Genres of *repente* that don't use *motes* may use *deixas* ("cues", "hooks", or "catchwords"). In these genres, the *repentista* must "take the cue" from the last verse line of the stanza just sung by the opponent, and make the first verse line of his/her reply stanza rhyme with it. For example, the *sextilha* is one of the easiest and most common genres of *repente*, and is typically the genre sung at the beginning of any *cantoria*. The stanza of a *sextilha* is composed of six verse lines, and the second, fourth and sixth lines must rhyme. The reply to a sextilha stanza must start with a line that rhymes with the stanza just sung by the opponent *repentista*. This creates a sense of continuity between the stanzas of different *repentistas*. Thus, because of the *deixa*, in a sequence of two stanzas of six lines, each sung by a different *repentista*, there will be rhymes on lines two, four, six, and seven (that is, the first of the second stanza). The *deixa* participates in creating the effect of dialogic improvisation between the *repentistas* (Sautchuk 2011, p. 27): because of the deixa, one *repentista* can manipulate the formal constraints imposed on the other *repentista*. The deixa also offloads verse composition: the *repentista* is provided with a "starting point", in contrast to the "end point" of the *mote*.

In a *repente* performance, information flows among the *repentistas*, their guitars, the pre-defined melodies they use, the versification structures they sing, and, of course, the audience. Thus, we can say, with Davies and Michaelian (2016), that they all constitute a DCS, exchanging information in order to solve the oral poetry improvisation problem. This problem is characterized by temporal, metric, thematic and dialogic constraints. The *viola*, the *toadas*, the *mote* and the *deixa* are part of the DCS, in the sense that they externalize and offload these constraints. This description follows a task-based notion of distributed cognition. It clarifies how a system of agents and artifacts can embed a formal structure of a cognitive task. However, from a Peircean perspective, what is still lacking in such a description is a more explicit account of sign action. The agency involved in the cognitive process of *repente* is not only the agency of individuals, with their motivations and choices, or the agency of a social group with its own motivations and choices, but also the agency of signs that beget more of themselves and tend towards stable growth of habits.

7.4 Repente as Semiosis

For active semiotic externalism, it is not enough to consider *repente* as a system of agents and artifacts. We have to take into consideration that these agents and artifacts instantiate a semiotic process. That is, cognitive agency has a semiotic ontology and it is the job of *semiosis* to try to explain what makes agents agents. The concept of semiosis is not a re-description of individual actions of agents and artifacts. In this view, whenever one refers to an agent or a cognitive artifact, he/she is referring to signs, and describing semiosis. This is the case for the *violas*, *toadas*, *motes*, *deixas*, *repentistas* and *audience* that are elements of a cantoria DCS. In what concerns the semiotic process of *repente*, they are all signs in action. This amounts to a *semiotization* of distributed cognition.

This semiotization represents a step further than Hutchins' (2011, 2014) characterization of distributed cognition as an emergentist and enculturated approach to externalist cognition. In our description, what emerges from—and downwardly organizes[5]—distributed cognitive systems is semiosis, a process with different properties than those of the system, which is nevertheless instantiated by the system. Semiosis exhibits properties such as triadic irreducibility (EP 2:171; CP 5.484; Brunning 1997; Burch 1997), processuality (CP 5.484; Atkin 2016, p. 132), irreversibility (CP 5.253, 5.421), continuity (MS 875; Parker 1998, pp. 75, 147; Jappy 2013, p. 19), tendency to the infinitum (EP 2:478–483; CP 2.92, 2.303; Atkin 2016, pp. 136–140), vagueness (CP 5.447), generality (CP 6.172; Potter 1997, p. 89), growth (EP 1:313, 2:10, 2:937). From this perspective, what defines and delimit the *cantoria* as a DCS are properties of a sign, *repente*, instantiated by the actions of agents and artifacts of DCS, but which, nevertheless, is irreducible to properties of those agents and artifacts. This is not to say that there is no task-relevant information flow, but rather, that what delimits and defines what counts as task relevant information flow is *repente*, a general sign-in-action which begets more of itself, which regulates agents and artifacts according to a tendency towards certain stable patterns of action.

Imagine the following scene: in a *cantoria*, a *repentista* plays his/her *viola* and sings verses that fit in a *mote* (or in a *deixa*) following a *toada*. What does one observe from this scene? In a task-based approach to distributed cognition, we observe a cognitive agent and cognitive artifacts—material objects with properties—exchanging information so as to perform a cognitive task. A Peircean cognitive semiotic approach produces a different description from the same scene: we observe the action of *repente* as a semiotic process through the action of the *repentistas*, *violas*, verses, *motes*, etc. *Repente* is a tri-relative pattern of action that delimits and constrains the cognitive behavior of agents coupled to artifacts.

Take a *viola* as a specific example. It is a material artefact with several different properties and affordances. But when we consider *repente*, we are not talking about properties and affordances of a *viola per se*, we are talking about the *viola* as a sign in action: the contextualized regularity of action embodied in a *viola,* in a particular history of uses of the *viola* and in the tendency towards regularity that this history determines. The usage of the *viola* in *repente* is not explained only by its material affordances plus representational states of a system working towards a problem-solving objective. One has to consider also a general habit of action, involving that artifact, that has grown across multiple situated instances in a certain historical trajectory. The *viola* solves functional problems, but as it does so it embodies *repente*. It involves a system of rules (for example, musical tuning), a grammar, a tradition of use and manipulation, a tradition of design and manufacture, etc.

To say that something is a sign in action is to say that it possesses a specific logical structure and hierarchy. This specific structure is a triadic logical relation that embeds a tendency towards regular behavior. In Peirce's description, this tendency develops through logically subsequent inferential stages: abductive, deductive, and

[5] The notion of "downward determination" (Queiroz and El-Hani 2012) refers to the constraining influence of higher- level boundary conditions over objects described at a meso- or focal- level of observation. In downward determination, higher-level boundary conditions coordinate or regulate the dynamics observed at the focal level.

inductive (Hookway 2002, p. 273; Forster 2011, p. 144). Abduction is the ampliative inference that begins with observation of a surprising outcome that doesn't fit into the habits and expectations of the observer and culminates with the formation and selection of a hypothesis as to how to subsume this outcome under regular patterns (Paavola 2012). It is like an "act of insight" that "comes to us like a flash" (CP 5.181), responsible for any "discovery", "insight", and is associated with creativity. Deduction is a non-ampliative inference that works out the experiential consequences of hypotheses, either necessary or statistical consequences (Forster 2011, pp. 141–144). Induction is an ampliative inference that tests the hypothesis formulated by abduction and explicated by deduction. Induction compares the results of deduction with results of experimental tests (Forster 2011, p. 147).

In Peirce's semiotics, inference is ubiquitously present in cognition, and is centrally related to the description of temporal distribution (W 2:240). It is because of continuous inferential steps that generalize the past and update it into the present that semiosis forms a trajectory, a temporal development that possesses the property of continuity—and not just a sequence of presents (CP 6.111). *Repente* embodies such an inferential structure: it is a semiotic process developing in time. A semiotized description of a cantoria DCS considers that this system enacts in time, through inferential relations, *repente* as a sign in action, negotiating its regularity in face of different local situations, specific events, or unusual and surprising outcomes. While a DCS can be scaled down into subtasks, a description of DCS as a semiotic process should also be able to break it down into inferential logical components. These are system-wide inferential processes that are logically (not necessarily chronologically) subsequent and whose result is to enact *repente* as a regular semiotic pattern.

- Abduction is the logical step that explains how the system adapts to new situations: for example, *repentistas* must respond to new themes, contexts and situations, but come up with ways to subsume these novelties into versified structures that can enact the regularity of effects of the *repente* sign. In this case, the cognitive artifacts work as abductive tools, helping to generate hypotheses of action according to anticipated outcomes.[6]
- Deduction is the logical step that works out the abductive responses according to the norms and rules involved in *repente*, checking if the abductive responses are sound or consistent according to these rules: for example, if a *repentista* first composes one verse line, now he/she has to consistently rhyme other verse lines accordingly, and, because of the *deixa,* also the opponent has to maintain consistency. In this case, the cognitive artifacts work by helping to constrain behavior towards a regular set of consequences.[7]
- Induction is the logical step that explains how the system evaluates the abductive and deductive steps so as to keep the stability of the *repente* sign: for example,

[6] Here, it is clear that there is a strong connection with the materiality of artifacts, with iconicity, and with the micro level of the emergence model, which we introduced in Chap. 2.

[7] Here, it is clear that there is a strong connection with constraints on artifacts, semiotic objecthood, and problem.

audiences react positively or negatively to innovations proposed and developed by the *repentistas*. In this case, the cognitive artifacts help the performance of experiments by facilitating perceptual judgements, such as when the *baião-de-viola* helps to compare and contrast response time of *repentistas*, or when the *toada* helps to identify metrical inconsistencies.[8]

In this description, inferential activity is embodied in the DCS as a generalizing tendency. The result of inferential activity is that a *cantoria* (a session of *repente*) is able to instantiate the sign *repente*. The system generates, develops, and evaluates/corrects meaning, so that it materializes and embodies *repente*. This is a semiotized reframing of a non-agent-centered notion of DCS: it points towards a sign-centered or semiotic-centered notion of DCS. DCSs are identified based on semiotic processes. What is most decisive here is that the DCS enacts a sign, and not exactly, as Davies and Michaelian (2016) argue, that the DCS solves a cognitive task.

7.5 How a Cognitive Semiotic Approach Addresses Some Questions About DCSs

We propose a semiotic externalist position, according to which a DCS is to be understood in terms of sign action. The DCS is delimited and defined by a sign in action, and the elements of the DCS are all understood as signs in action (see Table 7.1), as signs that generate interpretants. In this position, cognitive distribution is a necessary consequence of the nature of semiosis as a dynamic process, that only exists through actions, and that has to be distributed in time and space as a process materially instantiated through several organism-artifact interactions. Stability of semiosis in time is not understood as stasis (such as in static representational states) but as regularity of action.

Semiosis is a temporally distributed process in which a regular tendency towards certain future outcomes emerges from a history of sign interactions. Our approach here is pragmatic and action-oriented (cf. Engel et al. 2016; Madzia and Jung 2016), and focuses on sign action, as conceived by Peirce's pragmatism. We agree with Hutchins that cognitive externalism needs to be enculturated, and that this enculturation entails an emergentist position. But if the agent is not the factor that delimits and defines a distributed cognitive system, then what is? In our approach, what makes some entity a part of a DCS is a general sign, a habitual tendency in development, that the components of the DCS embody together. This sign is a general pattern of interpretant production, a tendency towards certain future outcomes which emerges from a history of interactions. Artifacts and agents embody sign action by participating in tri-relative actions that produce regular interpretative effects. In other words, what makes an entity a part of a DCS is that it acts as a sign so as to

[8] Here, there is a strong connection with the interpretant, with problem-solving, and with the habituation of problem-solving.

Table 7.1 A schematic comparison between different possible descriptions of the cognitive task of *repente*. In internalist and externalist descriptions, *repente* musicians process information and solve problems, either internally or externally, with the help of external artifacts. In a semiotic externalist position, problem solving is understood as semiosis

	Internalism	Externalism	Semiotic externalism
Repente	Internal problem-solving task	External problem-solving task	Semiosis (action of signs)
Action of the *repentista*	The *repentista* processes information through internal cognition	The *repentista* engages in information flow with external artifacts (e.g., guitar)	The *repentista* is a sign in action
Viola (*repente* guitar), verse, and other artifacts	An input and/or output of information processing, a representational state	A cognitive artifact that augments and distributes information flow	A sign in action
Agency	A property of individual agents (*repentistas*)	An emergent property, distributed among participants of a cognitive task	An emergent property of the action of signs
Audience	Parallel and synchronized information processors (individuals)	Distributed information flow	A sign in action (generating interpretants)

generate other signs. General sign action downwardly organizes artifacts and agents towards re-instantiations of itself in time and space. General signs thus delimit and define DCSs in terms of temporally distributed, habitual action. This view confers some sort of autonomous agency to the semiotic process itself, instead of locating agency only in the individual with his/her purposes and objectives, or yet in a society of individuals with their purposes and objectives.

One implication of our approach regards art as a metasemiotic investigation (as we will see in Chap. 10). Noë (2015) has described art as "strange tools" for philosophical investigation on the practices, techniques and technologies that organize us. In our description, *repente* amounts to a metasemiotic investigation. The improvisational dialogue of *repente* is a strange conversation, with a conventionalized beginning and end, rigid rules, regular repeating patterns, a winner and a loser. *Repente* enacts a highly constrained improvisational dialogue that submits natural language to strict and unusual conditions. It is a multi-level system of constraints: different levels—temporal, metric, rhymic, dialogical, situational—are constrained by each other, resulting in a system that continuously generates unusual situations and subsumes them according to a strange self-constructed regularity. This amounts to a metasemiotic investigation of the most pervasive cognitive artifact, natural language (Clark 2005), that provokes and negotiates surprising and regular semiotic behavior, participating in a process of niche construction (Sinha 2009; Magnani 2009).

The action of this sign is structured in the following way: the system submits language to unusual situations, in the form of an improvisational challenge; then, agents and artifacts offer solutions to the challenge, bringing to life a logical process that generates conjectures and possibilities (abductive step), develops them consistently with the rules of the problem-solving task (deductive step), and evaluates the results (inductive step) so as to form, or conform to, tendencies and regularities of action. In *repente*, semiosis subsumes surprising situations into habits, thus fueling the further development of *repente* as a sign. This power for a further development of interpretants, which become signs producing other interpretants—or, in other words, the capacity of signs of begetting more signs,—is a fundamental property that identifies cognition in Peirce's cognitive semiotics, and is for us the basis for a semiotic externalist approach to distributed cognition.

Chapter 8
Semiosis as Problem-Solving, Problem-Solving as Semiosis

8.1 Triad and Categories in Problem-Solving

We solve problems all the time throughout our lives. Problem-solving tasks are not just puzzles, such as the Tower of Hanoi, or wayfinding challenges, such as navigating the underground system. Creating and interpreting artworks, doing science and politics, running businesses can all be conceptualized as a series of complex and interlocked problem-solving tasks. Peirce's theory of inquiry can be conceptualized as having the form of problem-solving: the initial state is of doubt, the goal state is of belief, and the intermediate states are achieved by different kinds of inferential steps. In this chapter, we model semioses as problem-solving tasks. In doing that, we generalize the notion of a problem-solving task to include all sorts of activities that require habitualization of sign action.

The first step in modeling semiosis as problem-solving is to speculatively identify the fundamental notions of semiosis (sign, object, Interpretant, firstness, secondness, thirdness) in problem-solving tasks. In the previous chapters, we have already related the sign to artifacts in problem-solving (see Chap. 5), the object to problems in problem-solving (see Chap. 6), and interpretive inferential activity to the process of solving a problem (see Chap. 7). Following these suggestions, we can characterize problem-solving as a process involving artifacts, problems, and solutions as co-dependent participants of a triadic relation-in-action. An implication is that we differentiate three levels or dimensions involved in the activity of problem-solving—the artifact (sign, or S), the artifact in relation to the problem (sign-object, or S-O), the solution as the mediator of the relation between artifact and problem (sign-object-interpretant, or S-O-I). Traditional problem-solving models are unidimensional and tend to postulate phases or stages (framing, searching, etc.). In our approach, problem-solving is not treated as a succession of phases or stages, but as a hierarchical interaction—at the level of firstness, problem-solving is a process concerning the nature of artifacts; at the level of secondness, it is a process

P. Atã, J. Queiroz, *Mind as an External Sign: A Semiotic Externalist Philosophy of Cognitive Science*, Synthese Library 531,
https://doi.org/10.1007/978-3-032-08988-5_8

concerning the nature of the relationship between artifacts and the problems they embody; at the level of thirdness, it is a process concerning the nature of the interpretation of this relationship (between artifacts and problems). Following the triadic irreducibility of Peirce's model, our approach suggests operational irreducibility between the problem, the artifact, and the solution process. There is no problem dissociated from artifacts and from a solution process, and vice-versa. This is in line with situated problem-solving (Kirsh 2009), and its central idea that artifacts not only "refer to" an abstractly formulated problem, but actually embody the problem for an agent-in-action.

Problem-solving is a triadic process involving three indissociable elements: an artifact (1), which represents a problem (2), so as to allow for a solution (3). This triadic frame brings with it all the implication of Peirce's category system: firstness is associated with qualitative possibilities and with the sign, secondness is associated with reactive effects and with the objects, and thirdnesss is associated with habitual regularity and with the interpretant. Respectively, solving a problem involves three levels of semiosis: framing a qualitative problem by means of artifact qualities (1), searching through a reactive problem by means of actual artifact states (2), habitualizing the solution to a regular problem by means of artifact habits (3). These are not necessarily subsequent steps in chronological order. Their ordering reflects a logical presupposition: the third level involves the second, the second involves the first, and the first has some degree of autonomy.

The first level of sign action—firstness of problem-solving—is mainly dependent on the qualities of the artifact. The problem, at this level, can be characterized as qualitative, because it is iconically determined by the artifact. The solution process, at this level, can be characterized as framing. The artifact has the capacity to frame a problem, giving it a material body and providing the conditions so that manipulation of the artifact allows solution of the problem. Framing a problem involves setting up a task environment, recruiting material resources from the environment and arranging them in specific ways. The term artifact, here, may also include body parts and movements of the agents involved. The core of this operation is the engagement and manipulation of physical and perceptual qualities of artifacts (qualisigns) so as to tentatively define a problem, the qualitative problem (which is an icon) that hypothetically allows for a solution (framing, which is a rheme). This level of sign action has a certain degree of autonomy and independence, in the sense that explorations in the artifact may actually transform the nature of the problem and the nature of the solution, so that one may explore artifacts for their own sake, not guided by a theoretical understanding of a problem, but actually seeking for hypothetical ways to externalize a problem, that is, iconically determine a problem.

The second level of sign action—secondness of problem-solving—is mainly dependent on the reaction of the problem. By reaction of the problem, we mean the capacity of the problem to act back at the artifact that embodies it, compelling the artifact to assume some configurations instead of others, thus bringing the problem-solving from the realm of possibilities to the realm of existents. The notion of a reactive problem involves the semiotic operation of mapping problem states onto artifact states, especially the goal or final state, so that solving the problem becomes a matter of manipulating the artifact so as to achieve the goal state in the artifact. The solution process at this level corresponds to searching the problem space mapped onto the artifact states. At this level, the agent does not engage anymore the artifact as an artifact itself: he/she views the situated artifact as the problem, so that engagement with the artifact is viewed as an engagement with the problem, a search of a problem space indexically signified by artifact states. Here we are dealing with actual existent states of the artifact (sinsigns) rather than qualities alone.[1] This is not anymore a free manipulation of artifact qualities so as to hypothetically frame a qualitative problem, but rather a manipulation compelled by the necessities of a reactive problem (index), which realizes in actuality the solution process (dicent). The solution process, at this level, can be characterized as searching. Searching has a dicentic, or proposition-like, logical structure, in which artifact states are taken to be predicates of the current state of a problem.

The third level of sign action—thirdness of problem-solving—concentrates on the solution, and can be referred to as habitualization. A fully solved problem is one in which the solution is not only reached, but generalized for all habitual cases, that is, theoretically formalized (symbols), and understood (argument), so that the problem may always be solved under similar circumstances. This involves generalizing, from the situated task environment at hand, to any representation that frames that problem under similar conditions (legisign).

8.2 Trichotomies and Sign Classes

Identifying the semiotic triad and the categories of firstness, secondness and thirdness in problem-solving allows us to apply Peirce's model of 10 classes of signs (see Chap. 2, Sect. 2.7) to problem-solving. The objective of doing that is to unveil the complex structure of constraints involved in solving a problem. As explained in Sect. 2.7, the 10 classes of signs are derived from three trichotomies. Each trichotomy breaks down a component of the semiotic triad (sign, object and interpretant), into their phaneroscopic elements of firstness, secondness and thirdness. In problem-solving, this means breaking down the artifact (first trichotomy), the problem (second trichotomy) and the solution (third trichotomy) into a complex structure of interactions between qualities, reactions and habits (see Table 8.1). Following the premises of process philosophy, artifact, problem, and solution here should all be understood as processes. Considered together, the three trichotomies express how

[1] Of course, the artifacts do not need to be the same in each phase (defining the problem and solving it). It seems reasonable to assume that certain problems are defined/discovered through one artifact, and only later is a second artifact constructed to frame the problem and make it manipulable.

Table 8.1 The first trichotomy models the artifact, the second the problem, and the third the problem solution

first trichotomy	sign (S)	artifact	3—habitual artifacts (legisign) 2—existent artifacts (sinsign) 1—artifact qualities (qualisign)
second trichotomy	object (S-O)	problem	3—habitual problem (symbol) 2—reactive problem (index) 1—qualitative problem (icon)
third trichotomy	interpretant (S-O-I)	solution	3—habitualization (argument) 2—search (dicent) 1—framing (rheme)

artifact, problem, and solution constrain, synchronically and diachronically, each step of problem-solving.

Let's start with the first trichotomy. During problem-solving, the agent manipulates artifacts in a task environment. However, there are different aspects or natures that make up an artifact, and which the agent may interact with at different stages. On the level of firstness, the agent manipulates a set of material and sensorial qualities embodied in the artifact (shapes, colors, sounds, phonemes, relations, parallelisms, textures, etc). This is the artifact as a qualisign. On the level of secondness, the agent manipulates the artifact as a token, a singular existent in its specific situated environment ("this piece on the board", "this line on the map", "this pattern of rhyme" …). This is the artifact as a sinsign. On the level of thirdness, the agent interacts with the artifact as a set of distributed and general habits, a type (a kind of puzzle-game, a kind of map, a genre of improvisation, regardless of the specificities of this or that token). The artifact here is a general entity composed of a set of regularities. This is the artifact as a legisign. In short, we interpret qualisigns as artifact properties, sinsigns as existent artifacts, and legisigns as habitual artifacts. In problem-solving, these different levels of artifactuality constrain the solving activity in different ways.

Through the manipulation of the artifact, the agent engages with a problem. The second trichotomy (icon, index, symbol) helps to clarify what is the problem with which the agent engages. On the level of firstness, the problem is tentatively defined by the artifact, without a very clear indication of how exactly the states of the artifact correspond or not to states of the problem. At this level, the problem has the character of an icon, something that only possibly exists, but whose actual modes of operation haven't been discovered or defined. We call it a *qualitative problem*, a problem only qualitatively defined or framed. For example, we can approach urban navigation in terms of a geographically accurate map, which is one kind of artifact, or in terms of an diagram of relations like an electrical circuit diagram, which is another kind of artifact (for example, the subway map), and these are two different ways of iconically framing the problem of urban navigation. These different frames alter the space of possibilities on which the solution is to be developed, even before any actual attempt at solving the problem. On the level of secondness, the problem is experienced as a set of hard, external, constraints that compel the artifact. Here, problem states are not automatically embodied in the artifact, but rather mapped onto it, imposing restrictions on its properties. Following Peircean vocabulary, we can say that the problem is *reactive*, which is to say that it is actually available through the artifact by forcing this artifact to behave in some way. For example: a certain underground diagram is forced to display a certain structure of connection of stations and lines because that's how they really are connected in the city, regardless of whether this is convenient or not for the artifact. Here, the artifact does not determine the problem, but rather the problem determines the artifact. This is the problem as an index. Note that a reactive problem requires an actual, existent artifact (sinsign). On the level of thirdness, the problem is experienced as a general pattern, distributed across different possible representations. When I say "the problem of urban navigation", I am referring to such a pattern, a habit generalized from many instances of specific problem-solving processes. This is the problem as symbolically defined. In short, we consider icons as qualitative problems, indices as reactive problems, and symbols as habitual problems. In problem-solving, these are different levels of a problem, all of which interact and influence problem-solving.

The third trichotomy describes the interaction or engagement of the agent with the problem through the artifact. This is what we identified as the solution process. On the level of firstness, the solution process amounts to tentatively framing a problem through an artifact, which can just offer hypothetical possibilities of solution. This is a rhematic level of problem-solving. On the level of secondness, the solution process amounts to a mapping and searching in a problem through the artifact. Search differs from mere framing in that it has a target, typically a solution-state. In this mode of engagement, the agent explores the problem space in a targeted way, for example, by comparing actual problem states, as signified by the artifact, to a goal state. This is a dicisign level of problem-solving. On the level of thirdness, beyond searching for a solution state, the process consists in habitualizing the solution at hand—figuring out how a problem is constituted, how a solution works and under which circumstances will it continue to work. This is an argumentative level of problem-solving. In short, we consider rhemes as framing, dicents as searching a problem, and arguments as habitualizing a problem-solving process.

As previously explored, in Sect. 2.7, the three trichotomies, combined, form 10 classes of signs, and not 27 classes (3 × 3 × 3), due to how the hierarchical ordering of the trichotomies preclude most of the classes generated by trichotomic combination. For instance, artifact qualities can only stand to a qualitative problem—and not for a reactive or a habitual problem—for an operation of framing—and not for searching or for habitualization. The ten classes of signs described in problem-solving are shown in Table 8.2 below.

In short, there are three semiotic components in problem-solving—artifact, problem, and solution—and three kinds of constraints that each of these components can exert on problem-solving—qualitative, reactive, and habitual. Put together, they determine three trichotomies and 10 classes of signs in problem-solving. The categories model the different ways in which artifact, problem and solution constrain problem-solving. In the next section, we explore the interrelation between the 10

Table 8.2 10 sign classes interpreted as problem-solving process

333	Artifact habits, habitual problem and solution (habitualizing) Habitualizing the solution to a regular (habitual) problem by means artifact habits
332	Artifact habits, habitual problem and reactive solution (searching) Searching of a distributed problem space, through manipulation of artifact habits
331	Artifact habits, habitual problem and qualitative solution (framing) Framing of a distributed problem space, through manipulation of artifact habits
322	Artifact habits, reactive problem and reactive solution (searching) Search of a mapped problem space, through manipulation of artifact habits
321	Artifact habits, reactive problem and qualitative solution (framing) Framing a mapped problem space, through manipulation of artifact habits
311	Artifact habits, qualitative problem and qualitative solution (framing) Framing an embodied problem space, through manipulation of artifact habits
222	Artifact reactions, reactive problem and solution (searching) Search of a mapped problem space, through manipulation of actual artifact states
221	Artifact reactions, reactive problem and qualitative solution (framing) Framing a mapped problem space, through manipulation of actual artifact states
211	Artifact reactions, qualitative problem and solution (framing) Framing an embodied problem space, through manipulation of actual artifact states
111	Artifact qualities, qualitative problem and solution (framing) Framing an embodied problem space, through manipulation of artifact properties

classes, to arrive at a full picture of how the different constraints interact together to constitute problem-solving activities.

8.3 Relations Between 10 Classes: Balat's Diagram

During the 1980s, a group of semioticians from the University of Perpignan focused their efforts on studying the structures and relations present in Peirce's extended classifications of signs (10, 28, 66 classes of signs).[2] As a result, one of the participants of this group, Michel Balat, proposed three diagrams for the 10 classes of signs: a triangular diagram (Balat 1990, p. 81, Fig. 8.1), a square diagram (Balat 1990, p. 85), and a three-dimensional diagram (Balat 1990, p. 86). We will focus on one of them—the triangular diagram (Fig. 8.1).

The diagram shows the relationships (arrows) connecting each of the 10 sign classes. There are two kinds of arrows: pointing up and down. They correspond to two types of relationship that reduce all others: "involvement relationships" (2 > 1) and "instantiation relationships" (3 > 2). This generalization is based on two types of relationships—token>tone, 2 > 1; type > token, 3 > 2—and, in turn, are based on the relationships between the categories (CP 1.528, 5.49, 5.91) and describe how "generals" (types) "instantiate" an indeterminate number of "singulars" (tokens), or

[2] The main members of this group, under the influence and coordination of Gerard Delledale, were: Joelle Rethore, Antony Jappy, Robert Marty, and Michel Balat.

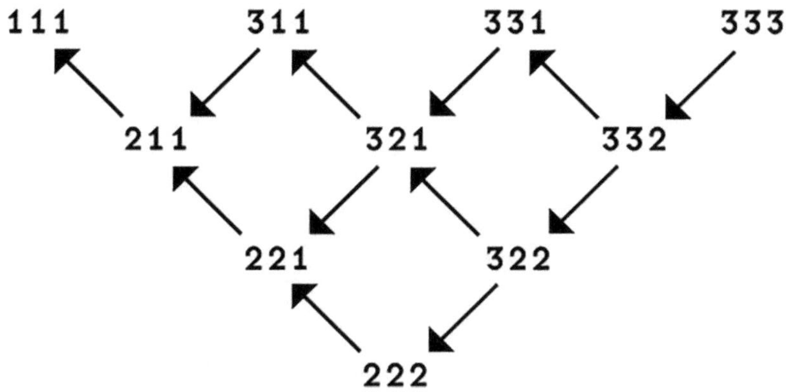

Fig. 8.1 M. Balat's triangular diagram for Peirce's 10 classes of signs. (Adapted from Balat 1990, p. 81)

existents, and prescribe their qualities (tone): "involvement" (secondness > firstness, 2 > 1), "instantiation" (thirdness>secondness, 3 > 2). Peirce asserted that 'laws' only become real through their instantiation in actual existents. Their reality is tied to the regularities we observe in nature. Simply put, laws don't exist on their own in some Platonic ideal; they only have being when realized in the concrete world (see Savan 1987–1988, p. 13). While laws are instantiated (existents), existents involve qualities (see Table 8.3). The diagram (figure XX) represents relations of involvement and instantiation between signs of decreasing generality, from right to left. Every arrow pointing down represents instantiation, and every arrow pointing up represents involvement. Thus, arguments (333) instantiate propositions (332) that involve rhematic symbols (terms, 331) to constitute the predicate of the proposition, and dicent indexical legisigns (322) to "indicate" its subject. Legisigns instantiate sinsigns, and so on down to the qualisign (111). Table 8.3, below, specifies each relation of instantiation and involvement shown on the diagram.

If we try to locate each of the trichotomies, which we discussed in the previous section, in Balat's diagram, we arrive at the results shown in Figs. 8.2, 8.3, and 8.4 below. The figures show how the properties of being habitual, reactive or qualitative, which are attributed to solutions, problems and artifacts, are distributed across a phenomenological space of semiotic relations. Figure 8.2 shows how artifact habits are ultimately instantiated by artifact reactions, which in turn ultimately involve artifacts qualities. Figure 8.3 shows how symbols are instantiated by indices which involve icons. Figure 8.4 shows how habitualization is instantiated by searching which involves framing.

We will use Balat's diagram here, to model the synchronic constraints acting in problem-solving. The diagram represents the complex structure of presuppositions and determinations between artifact, problem and solution in problem-solving. Each node in the diagram—each sign class—is a configuration of qualitative, reactive and/or habitual constraints exerted by the artifact, the problem and the solution. The arrows represent a hierarchy of containment and presupposition, so that 333

Table 8.3 Description of all the relationships of involvement and instantiation between each of the 10 classes of signs, as shown in Balat's diagram

Relation of instantiation	Relation of involvement
(333–332) Argument-Dicent instantiation; habitual solution instantiated in reactive solution; instantiation of habitual solution	(332–331) Dicent-Rheme involvement; reactive solution involves qualitative solution; involvement of a framing
(332–322) Symbol-Index instantiation; habitual problem instantiated in reactive problem; instantiation of habitual problem (when searching)	(322–321) Dicent-Rheme involvement; reactive solution involves qualitative solution; involvement of a framing (problem reactions to artifact habits)
(331–321) Symbol-index instantiation; habitual problem instantiated in reactive problem; instantiation of habitual problem (when framing)	(222–221) Dicent-Rheme involvement; reactive solution involves qualitative solution; involvement of a framing (problem reactions to artifact actions)
(322–222) Legisign-Sinsign instantiation; artifact habits are instantiated by artifact actions; instantiation of artifact habits (when searching a reactive problem)	(321–311) Index-Icon involvement; reactive problem involves qualitative problem; involvement of qualitative problem (by artifact habits)
(321–221) Legisign-Sinsign instantiation; artifact habits instantiated by artifact actions; instantiation of artifact habits (when framing a reactive problem)	(221–211) Index-Icon involvement; reactive problem involves qualitative problem; involvement of qualitative problems (by artifact actions)
(311–211) Legisign-Sinsign instantiation; artifact habits instantiated by artifact actions; instantiation of artifact habits (when framing a qualitative problem)	(211–111) Sinsign-Qualisign involvement; artifact reactions involve artifact qualities; involvement of artifact qualities

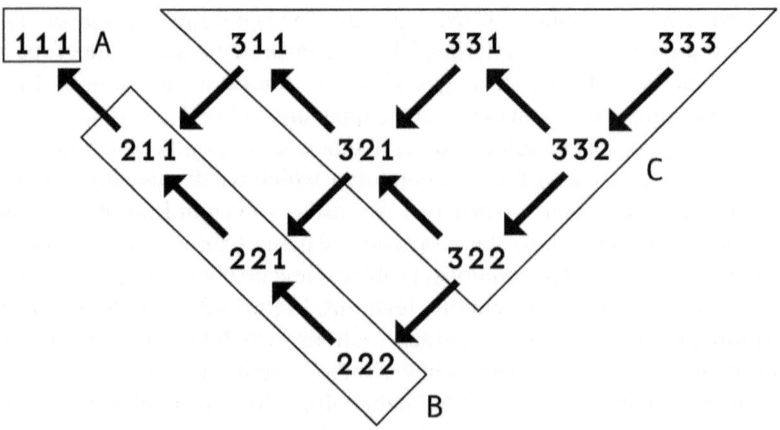

Fig. 8.2 Balat's diagram segmented according to the first trichotomy (artifacts). Area A corresponds to qualitative artifacts (qualisigns); Area B corresponds to reactive artifacts (sinsigns); Area C corresponds to habitual artifacts (legisigns)

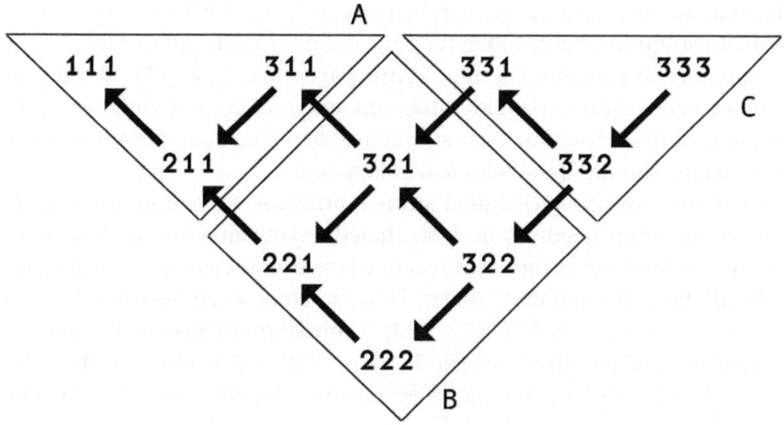

Fig. 8.3 Balat's diagram segmented according to the second trichotomy (problems). Area A corresponds to qualitative problems (icons); Area B corresponds to reactive problems (indices); Area C corresponds to habitual problems (symbols)

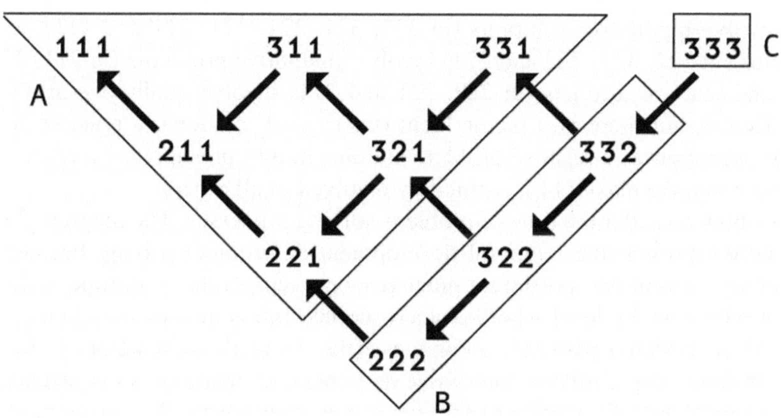

Fig. 8.4 Balat's diagram segmented according to the third trichotomy (solutions). Area A corresponds to qualitative solution, or framing (rhemes); Area B corresponds to reactive solution, or searching (dicents); Area C corresponds to habitual problems, or habitualization (arguments)

(the rightmost node, which represents a fully habitualized problem-solving process) presupposes and contains all the other nodes, while 111 (the leftmost node, which represents a fully qualitative problem-solving process) does not presuppose nor contain any other. How does the diagram represent a process of problem-solving? Let's consider two possible ways of reading the diagram as a representation of problem-solving—the mereology of problem-solving and the temporal succession of actions in problem-solving.

First let's consider a mereological approach to understand how the diagram represents a problem-solving process. From this perspective, the sign class of arguments (333), represents a complete process of solving a process, which can roughly be defined as finding a stable and habitual structure of constraints between artifact, problem and solution. We can then read the diagram from right to left as a mereological structure of problem-solving. Starting from the node 333, at the rightmost and topmost vertex of the triangular diagram, and moving left and down following the direction of the arrows, at each successive step we decompose this whole process of problem-solving into its logical components.

Each habitual constraint (habitual artifact, problem, and solution), or each number 3 in the diagram needs to be instantiated, rightwards and downwards in the diagram by a number of component reactive constraints (numbers 2). This is represented by all the arrows pointing down. Thus, habitual solutions (in 333) are instantiated by solution steps (in 332, 322, 222); habitual problems (in 331 and 332) are instantiated by reactive problems (in 331 and 332) and habitual artifacts (in 331, 321 and 311) need to be instantiated by reactive artifacts (in 211, 221 and 222). There is a hierarchical order in these chains of instantiations: habitual solutions require habitual problems and habitual problems require habitual artifacts, in this order. Note that the class 333 ultimately instantiated by all others. In turn, each reactive constraint (reactive artifact, problem and solution) necessarily involves a number of component qualitative constraints. Thus, reactive solutions (in 332, 322 and 222) involve qualitative solutions (in 331, 321, 221, 311, 211 and 111); reactive problems (in 322, 321, 222 and 221) involve qualitative problems (in 311, 211 and 111); and reactive artifacts (in 222, 221 and 211) involve qualitative artifacts (in 111). Here again there is a hierarchical ordering of relations: a reactive solution already presupposes a reactive problem and this in turn presupposes a reactive artifact. Note that the class 111 is ultimately involved in all others.

Let's now consider the time of problem-solving processes. The diagram does not accurately represent the temporal development of problem-solving because each level of sign action has a different phaneroscopic temporality—thirdness, the category of habits, is the level where phaneroscopic time is historical and cumulative, made of generalized patterns; secondness, the category of reaction, is the level where phaneroscopic time is successive and concrete, made of actual events; firstness, the category of qualities of feeling, is the level where phaneroscopic time is made of a vague and undefined present.

When we think of the successive steps or actions that an agent takes in a problem-solving tasks, we are talking about the time of secondness. If we look at the diagram, this time of successive reactive events is reflected in the class 222, which is the node at the bottom of the triangle diagram. Thus, if we want to grasp how the structure of constraints represented in the diagram acts in each successive step in problem-solving, we can focus on the class of dicent indexical sinsigns (222). In this case, the process of solving a problem can be understood as a temporal succession of dicent indexical sinsigns. Each of these actions taken by an agent, is subject to qualitative, reactive and habitual constraints stemming from the artifact, problem and solution, and this is what the diagram models.

If we move left from 222, we examine the qualitative aspects involved in each action taken in a problem-solving task. These qualitative aspects provide initiating conditions[3] for the focal action emerging at 222. A reactive solution (as in 222) involves and presupposes a reactive problem (as in 221, a rhematic indexical sinsign), but this in turn involves and presupposes a reactive artifact (as in 211, an iconic sinsign). Or, reading it the other way around, a qualitative artifact (111, a qualisign) provides the initiating conditions for a reactive artifact (211, an iconic sinsign), and this for a reactive problem (221, a rhematic indexical sinsign) and this in turn for a reactive solution (222, dicent indexical sinsign), which is an action taken, and we are back to the point where we began.

If we move right from 222, we examine the habits involved in each action taken in a problem-solving task. Each reactive step in problem-solving instantiates habits. These habits are a constraining factor which participate in selecting, out of many possible actions, that which is actually taken. There are three of these boundary constraints, in successive order: the constraint from habitual artifacts, from habitual problems and from habitual solutions. Thus, each 222 instantiates, first, the habits of the artifact being used, then the habits of the problem being approached through the artifact, and then the habits of the solution being developed.

At this point, there are still three classes which we haven't covered in our reading of the diagram: 331, 321 and 311. You may have noticed that these are the only classes in the entire diagram which mix 3s and 1s, that is, boundary conditions and initiating conditions. These classes display a kind of interaction which is implicit but invisible in the diagram of semiotic emergence discussed in Chap. 4 (see Figs. 8.5 and 8.6), the interaction between macro and micro level.

8.4 Conclusion and Implications

In the internal computationalist approach to problem-solving, exemplified by Herbert Simon's approach, cognition is viewed as the manipulation of internal symbols, much like a computational process running on a mental machine. From this perspective, problem-solving is essentially an internal search through a predefined space of possible solutions. The mind works by navigating this space step by step, as if following a set of preprogrammed rules, with the assumption that the solution lies within the boundaries of an internally represented model of the problem itself. The brain, in this view, is treated as a closed system, performing algorithmic operations on these internal representations to arrive at an answer. As an alternative to the internal computationalist view, situated perspectives to problem-solving (ex. Kirsh 2009), have emphasized the role of embodied interactions between agent and environment. There is a tendency, in situated approaches, to emphasize secondness, in the sense of dynamic action-reaction between agents and artifacts co-present in an environment.

---

[3] As we saw in Chap. 4, semiosis can be modeled as a three-level hierarchical structure in which Firstness corresponds to the micro-level that provides initiating conditions, Thirdnes to the macro-level that provides boundary conditions, and secondness to the focal-level.

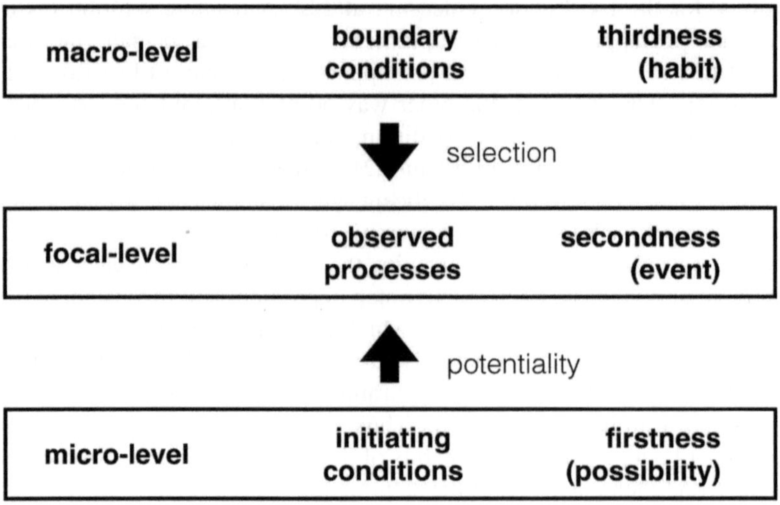

Fig. 8.5 Semiosis as an emergent multilevel process involving the determinative influence of initiating and boundary conditions, associated respectively to Firstness and Thirdness. These conditions regulate and select among the potential outcomes, influencing which possibilities are realized at the focal level

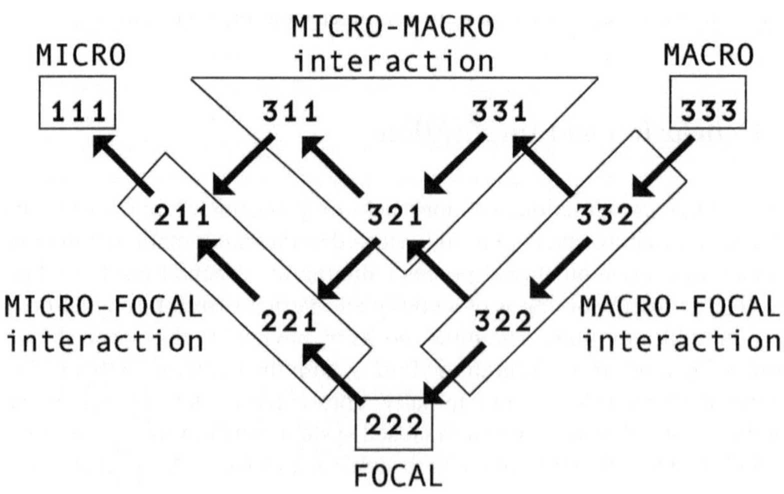

Fig. 8.6 A comparison between the multilevel model of semiotic emergence and Balat's diagram. Classes 111, 222 and 333 are treated, respectively, as the micro, focal and macro levels. Classes 211 and 221 correspond to the action of initiating conditions, from the micro-level to the focal-level, and which are mapped onto relations of involvement. Classes 332 and 322 correspond to the action of boundary conditions, from the macro-level to the focal-level, and which are mapped onto relations of instantiation

In dialogue with both of these approaches, but also in contrast to them, we propose to view problem-solving as a self-organized multilevel process of constraints operating between artifact, problem and solution, and based on a synchronic structure of sign classes which instantiate and involve other sign classes. The triangular diagram semiotically models this structure of many interdependent constraints. This diagram does not show a sequence of steps or temporal succession, and the temporality of problem-solving, in this approach, is not a temporality of linear successive states. Instead, the problem-solving process develops in time as a constant dynamic reproduction of itself as a self-organized process. Thus, the 333 class—habitual artifacts, problems and solutions—which ultimately involves all other classes, is, continuously, a product in process.

Chapter 9
Meaning Is Niche Construction: Semiotic Artifacts and Cognitive Externalism

How to provide a *locus* of observation for the formal notion of semiosis? The notions of niche and artifact are especially capable of updating the thesis, formulated by Peirce, that one cannot think without external signs, associating it to new empirical and theoretical methods and results. In this chapter, we introduce the notion of niche of semiotic artifacts. In our approach, cognition is semiosis, sign-action, in a process that takes the form of niche construction. In comparison with current usages of the term artifact to mean a material "thing" which is produced, semiotic artifacts are processes, signs-in-action. Niches of semiotic artifacts are structured spaces of fundamental conditions for stability of sign-action, conditions such as situatedness (co-localization) and temporal distribution between communities of agents, artifacts, and their environments. Niches of semiotic artifacts offer conditions for emergence of habit and surprise in semiosis/cognition. This line of inquiry suggests a cognitive semiotics framework based on dynamical, distributed, and emergent relations.

9.1 Where Do Signs Act?

Providing a *locus* for semiosis is a persistent problem in Peircean semiotics. In 1908, Peirce referred to an interpretant as an "effect on a person" (SS 80–81), an affirmation which suggests that the niche of semiosis is a person. But he admitted, in the same paragraph, that locating the interpretant to a person was a concession, a "sop to Cerberus," which he made so that readers could better grasp his theory. This suggestion, of the individual person as the niche where semiosis inhabits, conflicts with the perspective that Peirce's pragmaticism casts on the very idea of a person. Pragmaticism just cannot agree with the notion, common-sensical in modern times, of a person as an agent endowed with an individuality that separates them from their environment. The individualism that a person attributes to themselves is a "vulgar illusion of vanity" (CP 7.571), that overlooks synechism, the doctrine of continuity. In synechism, the philosophy that abhors neat separations between one and other, a person is not entirely identical to themselves, and not wholly distinct from other people around. On another occasion, Peirce states that "the human social circle

P. Atã, J. Queiroz, *Mind as an External Sign: A Semiotic Externalist Philosophy of Cognitive Science*, Synthese Library 531,
https://doi.org/10.1007/978-3-032-08988-5_9

(however broadly or narrowly that phrase may be understood) is a sort of vaguely compact person, in some aspects of superior position to the person of an individual organism" (EP 2: 338).

One of the main thesis in this book is that Peirce can be considered a precursor to cognitive externalism and the theses of situatedness and distributed cognition (Atã and Queiroz 2013, 2014, 2015, 2016, 2019; Atã et al. 2014). Peirce's semiotics does not isolate cognitive and semiotic processes into separate domains and subdomains (W 1:173). The thesis "cognition is semiosis," or "action of signs," emerges in Peirce's work associated with the construction of an anti-Cartesian philosophical program, presented in a series of articles published between 1868 and 1869 in the Journal of Speculative Philosophy (W 2:211–242; Bernstein 2010, p. 145). This thesis integrates cognition, situated experience, action of signs, and reality within a single theoretical framework. In Active Semiotic Externalism, mind, semiosis, and the manipulation of situated artifacts are intimately interconnected. Semiosis is not confined to the individual agent, within an individual mind or brain, but rather exists in communities of agents and in a universe imbued with signs-in-action. It is along these lines that we should conceptualize its niche. This line of research suggests a cognitive semiotics based on environmental, dynamic, distributed, and emergent relationships, in contrast to any cognitive semiotics centered on individual agents and static, autonomous units of meaning, where agents act as tutors or controllers of sign action.

In this chapter we examine different notions of niches, and propose two main ideas: first, that semiosis develops in niches of semiotic artifacts; second, that semiosis can be defined as cognitive/semiotic niche construction. In the following pages, we introduce the notion of niche in three versions—ecological, cognitive, and semiotic. We then reinterpret the concept of cognitive artifact in light of Peirce's process semiotics. Finally, we suggest the notion of niches of semiotic artifacts and explore some implications.

9.2 Niches

9.2.1 Ecological Niche

The word "niche" evokes the image of an architectural recess where a statue is displayed, a negative space that can be occupied, shaped, and where an element fulfills a function. In ecology, the concept of ecological niche describes the combinations of environmental factors that allow species to exist in a certain physical region or biotic community, as well as the effects that species have on these environmental factors (Peterson 2011, p. 14). In its classical definition by G. Evelyn Hutchinson, an ecological niche is an n-dimensional hypervolume, with each dimension representing significant ecological factors for the success of a species (Hutchinson 1957).

The concept resembles the ecological context of an organism, the surrounding environmental conditions that make its existence possible and probable.

There is a duality in understanding the ecological niche, as it can be defined as the "address" of an organism—where it typically thrives—as well as its "profession"—what it typically does in order to thrive (Hoffmeyer 2008). This duality illustrates an important characteristic of the ecological niche concept, which can be generalized to any niche concept—it integrates situatedness (the co-location between the agent and the environment) and stability of action (the typical activities of an agent in this environment), challenging dualistic separations between activity and material space, and between agent and environment.

The notion of an ecological niche allows for close ontological and bidirectional interdependence between organisms and their environment. According to Lewontin and Levins (1997), there is no organism without an environment, and vice versa, no environment without organisms. The construction of an ecological niche is a process of coevolution between organisms and their environments. The Theory of Niche Construction (Odling-Smee et al. 2003; Scott-Phillips et al. 2013) is based on the idea that organisms transform environments, and environmental processes, in turn, shape organisms.

On the one hand, every animal's behavior actively shapes its environment and potentially transforms it. This includes both large-scale environmental transformations with visible effects (e.g., beavers constructing dams and altering entire landscapes and ecosystems) and subtle cases of environmental change (e.g., warm-blooded animals constantly creating a thin layer of warm and humid air around themselves—their individually maintained micro-environment [Lewontin and Levins 1997, p. 97]). The environment accumulates the results of organisms' behaviors, and environmental transformations are passed down to future generations. This constitutes a system of inheritance that differs from genetic inheritance. On the other hand, changes in the environment transform the selective pressures on organisms and alter the n-dimensional hypervolume of ecological factors that defines the patterns of typical activity for an organism. As the environment changes, it shapes the characteristics and behaviors of organisms, leading to further coevolutionary processes.

This understanding of ecological niche construction highlights the dynamic and reciprocal relationship between organisms and their environments. It emphasizes the active role of organisms in shaping their surroundings and how these transformed environments, in turn, influence the adaptive pressures on organisms. This coevolutionary perspective is crucial for comprehending the complexity of ecosystems, species interactions, and the ongoing process of adaptation and evolution in the natural world.

Organisms and environments coevolve as emergent outcomes of a self-organized process of niche construction. The concepts of niche and niche construction emphasize how organisms and environments are interdependent, both from a dynamic perspective, as mutually interrelated parts of an evolutionary process, and from a situated perspective, always necessarily co-located.

9.2.2 Semiotic Niche

Jesper Hoffmeyer (2008) criticized the de-semiotization of the concept of ecological niche. According to him, it is not enough to merely describe the set of ecological factors on which an organism depends. It is essential to consider how the organism, within the dynamic process of niche construction, interprets its environment. The ecological factors that constitute the organism's niche, in this case, are not merely limiting factors for its well-being. They are actively interpreted by organisms. This reframing can be seen as a semiotization of the notion of ecological niche. Hoffmeyer used the term "semiotic niche" to refer to the world of signs (cues) in the environment of animals (or species) that they need to interpret to live fully (Hoffmeyer 2008, p. 7).

The semiotic relationship between organisms and their environment is bidirectional. On one hand, the organism must be capable of interpreting information from the environment; on the other hand, ecological factors relevant to the organism must provide cues that the organism is capable of interpreting. Hoffmeyer considers the semiotic niche as the external counterpart of the concept of "Umwelt". While Umwelt refers to how an organism perceives its environment, the semiotic niche pertains to the information that an environment can provide to the organism. The concept of the semiotic niche situates the organism in its environment within a semiotic dimension.

9.2.3 Cognitive Niche

Research on the evolution of cognition (Tooby and DeVore 1987), culture (Laland et al. 2000), and language (Bickerton 2009) has allowed the concepts of niche and niche construction to be "adapted" from ecology and evolution to the philosophy of cognitive science (Clark 2005). Tooby and DeVore (1987) were the first to propose the term "cognitive niche": a specific ecological niche that our human ancestors would have constructed at some point in our evolutionary history, which would explain a series of characteristics of our species arising from a common origin, such as complex organized behavior, language use, cultural learning and transmission, and division of labor. Pinker (2010) refined and expanded the concept to address "the coevolution of cognition, language, and sociality." The concept of the cognitive niche underwent a significant shift in usage with Clark (2005). While Pinker considered language as an adaptation of the cognitive niche, Clark approached language as the cognitive niche itself:

> Language, I have tried to show, is usefully understood as an animal-built material structure (a cognitive niche) that systematically alters the computational burdens involved in learning, reasoning and self-control. In this respect language stands to thought as a self-constructed behavior-enhancing niche stands to its animal "occupant." (Clark 2005, pp. 264–265)

In Clark's writings, the cognitive niche is related to the materiality of artifacts and structures in which cognition is distributed. This material understanding of the cognitive niche aligns with the author's externalist view on extended mind (Clark and Chalmers 1998). According to Clark, humans are cognitive cyborgs, symbionts "whose minds, and themselves, are distributed across biological and non-biological circuits" (Clark 2004, p. 3). Clark defines the construction of the cognitive niche as:

> … the process by which animals build physical structures that transform problem spaces in ways that aid (or sometimes impede) thinking and reasoning about some target domain or domains. These physical structures combine with appropriate culturally transmitted practices to enhance problem solving and, in the most dramatic cases, to make possible whole new forms of thought and reason.

A different perspective on cognitive niche construction is presented by Lorenzo Magnani. According to him, humans are 'chance seekers' (Magnani 2007; Bardone 2011), continuously engaged in a process of constructing and extracting latent possibilities to discover valuable information, novelty, and knowledge (Magnani 2007, p. 918). He emphasizes cognitive niche construction as a process through which we create affordances. Cognitive niches are defined as sets of possibilities, described as the human exploration of external resources and the incorporation of these resources into cognitive systems (Magnani 2009, p. 332).

9.3 Artifacts, Niches, and Habits

In this book, we have previously treated signs (the first element of the triad sign-object-interpretant) as artifacts, which embody problems (objects), in order to allow for attempted solutions to problems (Interpretants). In this perspective, the thesis that "all thought is in signs" can also be understood as "all thought is embodied in artifacts." This proposition requires that we reexamine the notion of artifact. Cognitive artifacts are material structures that distribute and externalize cognitive processes (Norman 1993; Hutchins 1995a, 2014). However, we should be careful with how we understand the notion of artifact, and not confuse an artifact with a thing. Based on an epistemology and an ontology of processes, an artifact is not a thing, a static substance, as, say, in a pair of glasses when we consider its material composition, but the capacity of that thing to participate in a process; thus, a pair of glasses when we consider its capacity to rest in front of the eyes and to organize light in a way that suits the necessities of the person wearing it, that is, the glass is only an artifact insofar as it is an extension of the process of vision, and thus dynamic. In this capacity for extension of a process rests its "glassness", its artifactual nature, and not in any of its constitutive substances.

There is an interconnection between an artifact and a niche, in the sense that both regard a "profession" together with an "address": a typical pattern of action embodied in a material environment. The pair of glasses is a pattern of action, an organization of vision, and thus it can only "live and thrive" as such in the environment of vision: in front of a person's open eyes. Clark's notion of cognitive niche takes advantage of this similarity of artifact and niche and makes it explicit: language, this quintessential cognitive artifact, is also a cognitive niche that allows for a number of cognitive processes that would be impossible otherwise.

Artifacts and niches are the central participants in a complex game of stability and chance. On the one hand, the niche is the negative space that provides stable conditions for semiosis/cognition to develop—it is because the niche of language is there in the world, reliably available, that several cognitive processes may develop and act predictably. On the other hand, the artifact, especially the icon, transforms problems and creates new problems, putting tension on those niche conditions and pushing for change in semiotic/cognitive abilities.

In the Peircean terminology, a notion that captures this game of stability and change is the notion of habit (Atã and Queiroz 2016; Määttänen 2010). A habit is a stable pattern of action, a "pattern of constraints," and can take the form of a "conditional proposition" in which certain things should happen under specific circumstances (EP 2.388), like a "rule of functioning" (CP 5.397, CP 2.643), a disposition to act in a certain way under specific circumstances, or simply the "permanence of some relation" (CP 1.415). A habit is a regularity that possesses some degree of stability.

The notion is similar to that of a niche, for example when we consider that the ecological niche of a species can be expressed in terms of what that species typically does (i.e., the niche of burrowing). Similarly, a habit is a typical pattern of sign action—the sign of a red traffic light typically produces the interpretant of cars stopping in front of it. In this example, the traffic light (a symbol) is but an artifact that conventionally signifies the habitual action; for the artifact to be effective, it has to occupy a negative space, a "context", where the action of stopping the car already exists as a reasonable consequence of a sign. A red traffic light may occupy that niche, but also the raised hand of a traffic guard, or an unraised traffic barrier gate. The symbolic character of all these signs implies that the niche, the presupposed habitual action of stopping the car in certain situations of traffic—is more decisive in determining the effects of the sign than the material qualities of the artifacts involved. A symbol is an example of highly stable, habitually-regulated semiosis.

In Peircean philosophy, the acquisition of stable regularities is described as a process of taking habits, which is probabilistic and cumulative:

> … all things have a tendency to take habits. For atoms and their parts, molecules and groups of molecules, and in short every conceivable real object, there is a greater probability of acting as on a former like occasion than otherwise. This tendency itself constitutes a regularity, and is continually on the increase. In looking back into the past we are looking toward periods when it was a less and less decided tendency. But its own essential nature is to grow. It is a generalizing tendency; it causes actions in the future to follow some generalization of past actions; and this tendency is itself something capable of similar generalizations; and thus, it is self generative. (EP 1: 277)

At the core of the notion of semiosis, there is a dynamic interplay between change and stability through the accumulation of self-generated probabilistic regularities. For Peirce, "what something means is simply the habits that something involves"

(CP 5.400). A direct consequence of this is that Peirce's semiotics does not focus on the "sign" as a substantial unit, but on the "semiosis," the process of "action of the sign" (Fisch 1986, p. 330). Therefore, "meaning" is not a property of the sign itself (nor of the sign user) but of the habitual action of the sign in its niche. A semiotic artifact is always observed as an artifact-in-action in a situated niche. In this perspective, cognition takes the form of developing semiotic artifacts. Semiotic processes acquire habits of material action and stabilize them through the availability of material artifacts in an environment, not as substances that incorporate the semiotic process, but as the process itself in its materially stabilized form. We can treat cognition as the construction of niches of semiotic artifacts, where the notion of artifact refers to semiotic processes stabilized and materially instantiated, and the notion of niche refers to how artifacts acquire and lose stable habits of action depending on the role they play in an artifact-rich environment.

The notion of habit helps to characterize how the action of signs (development of semiotic artifacts) unfolds over time, acquiring, losing, and transforming dispositions and regularities. Semiosis (and habit) develop as dynamic trajectories of generalizations from past interactions and predictions about their future outcomes. Dispositions for action are situated and relational, requiring circumstances for action (Määttänen 2010). The situationality of an agent or artifact is inseparable from the temporal distribution of a chain of semiotic interactions. Semiosis is part of a universe already infused with signs, which evolves (Hausman 1993). Just as an ecological niche is characterized as a hypervolume of ecological and environmental conditions for the success of an organism or species, the notion of niche of semiotic artifact allows us to characterize the conditions for the action of signs (semiosis) to develop stably.

9.4 Where Is the Meaning? Niches of Semiotic Artifacts

In the opening of this chapter we asked a question: where do signs act? Let's consider a specific application of this question: Where do signs act in music? Or, to put in other words—where is the meaning of a musical passage? Is it inside someone's head? The composer's? The listener's? Is it in the sound-producing properties of musical instruments? Is it in the cultural milieu surrounding music production and consumption? If our argumentation in this chapter is correct, we should be able to indicate the meaning of music is a property of a niche of semiotic artifacts. This means that we should consider semiotic artifacts (treated here as processes and not substances) and the niche these artifacts fit into, which provides conditions for habitual sign action around music composition and consumption. Our focus on this section is on the semiotic *locus* of musical meaning and not its technical functioning. Here we briefly introduce examples of concepts used to explain musical meaning and how they can be understood as semiotic artifacts which fit into niches.

Investigations of musical meaning have used notions such as schemas, patterns, templates and conceptual metaphors to account for how heard sounds are framed as meaningful structures. Kendall (2010) stresses the fact that listeners make associations, and that these can be related to typical patterns or schemas. These schemas are central to the effort of listeners to discern meaning (Kendall 2010, pp. 63–64). An example is how listeners are able to discern musical events. According to Kendall, listeners make use of an "event schema" that helps them not only to listen in terms of events, but also to access past experiences in terms of a history of events. For example, continuous felt experience of energy flow dynamics in a musical passage can be discerned and remembered as musical events and be ascribed general labels such as "rough, bumpy, grainy, smooth or flowing" (Kendall 2010, p. 66). Note that such labeling rests on the assumption that adjectives commonly used to describe qualities of texture can be used to describe qualities of sound. We refer to this assumption here as the "texture metaphor". Lakoff and Johnson (1980/2003, p. 3), stressed the importance of conceptual metaphors for human cognition: "our ordinary conceptual system, in terms of which we both think and act, is fundamentally metaphorical in nature". Conceptual metaphors operationalize a target domain in terms of a source domain. Johnson and Larson (2003) apply the Theory of Conceptual Metaphors to the cognition of musical motion, arguing that musical motion is metaphorically conceptualized as physical motion:

> Our claim is that people have no robust way of conceptualizing musical motion without metaphor and that all reasoning about musical motion and musical space inherits the internal logic of such metaphors. If this claim is correct, and if the source domain for musical motion is motion in space, then the ways we learn about space and physical motion should be crucial to how we experience and think about musical motion. (Johnson and Larson 2003, p. 68)

The authors consider three ways through which we experience and learn about motion: (a) we see objects move; (b) we move our bodies; (c) we feel our bodies being moved by forces (Johnson and Larson 2003, p. 68). These three ways give rise to three main metaphors to conceptualize musical motion: the "moving music" metaphor, the "musical landscape" metaphor and the "music as a moving force" metaphor.

The moving music metaphor describes musical events as objects that move past a stationary hearer from front (future events) to back (past events). Examples given by the authors include expressions such as "Here comes the recapitulation", "The strings slow down now", and "The music goes faster here" (p. 69). The musical landscape metaphor conceptualizes music as a three-dimensional environment through which the hearer moves. Future events are the landscape ahead and past events are the landscape already crossed. This metaphor accounts for expressions such as "We are coming to the coda", "Let's see, where are we in the second movement?", "The melody rises up ahead" (p. 71). The metaphor of music as a moving force is based on the experience of being physically displaced by substances and processes such as water and wind currents or large objects. In the metaphor, music becomes the substance that acts upon the hearer. This metaphor accounts for conceptualizations of music as something that "blow you away", "carry you along", "take you on a roller coaster ride", or make you "swing" (p. 75).

The conceptual metaphors—music as texture, as something that moves, as a landscape, or as a moving force—are semiotic artifacts. They extend our capacity to listen to music, by taking advantage of material, external, properties such as textures, dynamic spaces, and the sense of weight and resistance of outside forces that compel our bodies. As semiotic artifacts, their role is to frame a problem, which is the interpretative challenge that a musical passage poses—What does it mean? What is it trying to do? How can I relate to it? The semiotic artifacts provide tools to work on this problem. They frame the problem in terms of easier, more direct and embodied questions—What is the texture of this passage? How does this passage move? What does it do to my body? How do I travel through it? Finally, they also suggest possible answers—it is smooth, or bumpy, or grainy… it goes up, or down, or zig zags around… it makes me accelerate, or brake, or stop for a while… it lifts me, or throws me down, or spins me round… Now that I possess all of these possibilities, I can make comparisons between passages, or musical pieces, or even composers and how they have different qualities, how I like or dislike these qualities, or what I feel or think when listening to them. But when I say that *I* am doing that, what I really should be saying is that the artifacts are doing that, and that my interpretation of the musical passages have been profoundly molded by the way the interpretative problem has been framed. By framing the interpretative problem in terms of the "music as a moving force" semiotic artifact, my listening has already fit into a space of possible answers, a niche. When I produce the interpretant "the music carries me along", I am merely navigating this conceptual space: moving forces may carry you along, so that's an answer that is likely to come up at some point if you frame the problem this way. The semiotic artifacts involved also set some of the limits and boundaries of this space of tendencies and likelihoods—if I am experiencing some musical passage strictly in terms of spatial movement my interpretation will be bound to this conceptual metaphor. No rough or smooth texture in this case.

The notion of niche of semiotic artifacts refers to a structured space of fundamental conditions for the stability of semiosis, which are given by situated use of these artifacts. If semiotic artifacts correspond to semiosis stabilized in material instantiations, niches of semiotic artifacts refer to the space of likelihoods and tendencies which underscore such stability. Conceptual spaces (Boden 1999), media, art schools and movements, technical and practical contexts, disciplines and domains of knowledge can be described as niches of semiotic artifacts. They are self-organized processes that confer conditions of stability and change for semiosis. Niches of semiotic artifacts: (i) enable stable semiotic behavior of artifacts, which may take and/or lose habits, (ii) explain how semiotic artifacts acquire conditions to maintain their stability in the face of disruptions, (iii) create logical conditions (inductive frequency) in an environment for predictive inferences about the future behavior of semiotic artifacts, that is, niches allow the occurrence of expectations about the production of interpretants, in the sense that interpretants are expected to fit the same niche, (iv) create normative conditions to characterize interpretants as being within or outside of expectations, that is, niches allow certain artifacts and their forms of behavior to be experienced as familiar or surprising regarding that niche.

Put differently, niches of semiotic artifacts offer initial conditions and boundary conditions for the emergence of habitual semiosis (materially stable artifacts that behave predictably) and, consequently, surprises (breaks of habit in the behavior of semiotic artifacts). Semiosis is distributed among various artifacts (and agents) as it develops through the accumulation and self-organization of stable patterns of activity among distinct material instances. This habitual behavior is dependent on a process of coevolution between semiotic artifacts and problems in their situated environments—the construction of niches of semiotic artifacts—through which artifacts acquire, develop, and eventually lose the capacity to act in a stable manner.

9.5 Semiosis Is Niche Construction

By proposing niches of semiotic artifacts as the *locus* of semiosis, we update Peirce's "sop to Cerberus" with important advantages. One of them is that the more recent debate on cognitive externalism is guided by a robust scientific literature in terms of empirical evidence (Clark 2008), based on the development of experimental methods in cognitive anthropology (Hutchins 1995a), experimental psychology (Kirsh 2009), biorobotics (Laschi and Mazzolai 2016), mathematics (Stjernfelt and Pantsar 2023; Vold and Dirk 2020), and dynamic systems modeling (Chemero 2009; Wheeler 2005). Another advantage is that the notion of niche construction also draws on a literature, especially in philosophy of biology, about evolutionary processes, centered on the idea of ecological niche construction and cultural evolution (Odling-Smee et al. 2003; Laland 2017).

A niche construction approach to semiotics updates Peirce's theory of mind, emphasizing its situational, distributed, and dynamic aspects. The notions of niche and artifact, applied to distributed cognition, function as an empirical-theoretical update of Peirce's anti-Cartesian thesis, formulated about 150 years ago, that one cannot think without external signs (W 2:211–242). Our approach provides a processualist direction for cognitive semiotics—semiotic/cognitive processes occur in niches of semiotic artifacts, not in the "minds" of agents. Locating semiosis in niches of semiotic artifacts (rather than in individual agents) is equivalent to stating that cognition is the development of external semiotic artifacts. We no longer characterize "thinking" as an individual activity, internal to the agent, and we do not separate "thinking" from "acting." When we assert that cognition, or semiosis, is not located in the individual agent, we are suggesting that the notion of individual agent is not the most appropriate observational scale to locate interpretants and explain how they are produced. Thinking requires situatedness and temporal distribution—"purely internal" thinking (manipulation of mental models [Johnson-Laird 1980]) is contingent and only one possible stage in a process of signs embodied in the form of artifacts in action, and therefore external and communal.

Thinking with external artifacts creates conditions for these artifacts to develop over time, updating the frequency of their habits of action, maintaining habits of action, or subjecting these habits to situations of surprise and transformation. Just as an animal's behavior modifies its environment in some way over time and for future generations, any use of a semiotic artifact modifies, even imperceptibly, the availability of expectations and material regularities instantiated for future interpretants. Semiosis is the construction of niches of semiotic artifacts. A semiotic process is built through constraints, frequencies, possibilities, and contingencies that are co-located and temporally distributed in environments filled with artifacts. At the same time, semiotic processes construct these environments, maintaining their stability or transforming them. To the extent that every sign action produces interpretants that modify the space of likelihoods and tendencies for production of further interpretants, every sign action is an act of niche construction.

The notion of a niche of semiotic artifact should provide new research problems related to the opening, stabilization, and transformation of niches, including a new approach to properties, relations, and dynamics of niches such as robustness, resource availability, and resource partitioning in niches of semiotic artifacts. A research program in cognitive semiotics that examines niches of semiotic artifacts as the *locus* of cognition/semiosis should be able to integrate propositions and theses formulated in the domain of philosophical speculation with theoretical and empirical advances in cognitive science (situated and distributed cognition) and evolution (niche construction theory). This entails formulating a pragmatist research program in philosophy of mind and cognitive science, aligned with the notion of niche construction, and inaugurated by Peirce's thesis of "mind as semiosis"—mind externalized and temporally distributed in the action of signs. Our approach should integrate what has been called the "pragmatic turn" in cognitive science (Engel et al. 2016). The notion of niches of semiotic artifacts should participate in the explanation of how semiosis acquires and maintains stability of action.

Chapter 10
Artwork Authorship as a Legisign-in-Action

Active semiotic externalism challenges the idea that semiosis is located in the minds of individual agents. In the last chapter, we have suggested that semiosis is located in niches of semiotic artifacts. In this case, semiosis is a self-organized emergent process of niche construction, and thus not tutored by individual agents. If this suggestion is correct, we need to reconceive the role of individual agents in semiosis. How do individual agents participate in semiosis? One scenario fit for the examination of this question is artwork authorship. Historically, artistic creativity has been one of the domains of human cognition most strongly associated with individual agency and genius, an association institutionalized in the notion of authorship as the intentional and tutored bringing forth of artworks by an individual or a group of individuals. Thus, in this chapter we examine the question: what is the impact of the externalist semiotic thesis on the notion of authorship in arts?

To examine authorship, we start with the notion of artifact. An artwork is an artifact, and authorship is a relation that takes place between artifact and agent. Sticking with process philosophy, we conceptualize artifacts as processes, in contrast with the idea that artifacts are to be understood as substances. The idea of artifacts as substances is an important element for an agent-centric view of authorship, because it separates the process of creation from the artifact as a finished product. In this view, an artifact is an intended product of human action. By viewing artifacts as processes, we no longer separate between a process that is active and creative (and located in the mind of an individual agent) and a product that is passive and created (and located in the external environment). Artworks are artifacts, and thus are active and self-organized. They are active participants in their own creation, and not passive objects of it. What about artists? In our approach, the notion of authorship has to be reconceptualized—as the artifact, the author is also a self-organized component in a creation process that determines the artwork together with the author. An author is not, rigorously speaking, an individual agent *qua* individual agent, and authorship is not a property or capacity of an individual agent:

P. Atã, J. Queiroz, *Mind as an External Sign: A Semiotic Externalist Philosophy of Cognitive Science*, Synthese Library 531,
https://doi.org/10.1007/978-3-032-08988-5_10

authorship is an emergent and self-organized sign that is predicated on a agent and downwardly determines the agency of that agent. In the last part of this chapter we examine the different parts of this sign, taking advantage of Peirce's ten classes.

10.1 Artifacts and Agents

Works of art have been classified as artifacts by a number of philosophers, including Marcia Eaton (1969), Gary Iseminger (1973), Denis Dutton (1979), Risto Hilpinen (1992, 1993), Amie Thomasson (1999, 2004, 2005), and Jerrold Levinson (2007). We center our attention on Hilpinen's theory of artifacts, which has been pioneering work in the philosophy of artifacts. For Hilpinen, artifacts are products of agents' actions:

> I take artifacts (in the strict sense of the word) to be physical objects which have been manufactured for a certain purpose or intentionally modified for a certain purpose. Various cultural objects are created by making artifacts; for example, we can say that artists create works of art by making certain artifacts. (Hilpinen 1992, p. 58, see also Hilpinen 1992, p. 59)

Hilpinen later extended the notion of artifact to include "belief systems" (Hilpinen 1995). What is a product of an action?

> Some of the objects which people cause to exist are artificial and cultural objects, and some of these objects are artifacts. For example, when people build a house, they produce the house and its parts, as well as a lot of scrap which is used for nothing. This example suggests a distinction between two kinds of artificial or cultural objects: Some of these objects are produced intentionally; they are intended products of human activity, whereas others are produced unintentionally. I propose to apply the expression "artifact" only to the intended products of actions. (Hilpinen 1992, p. 60)

Something is an artifact if it results from the intentional action of an agent.[1] For Hilpinen, the agent is attached to the result (the artifact) as its cause (the agent's causation, or group of agents)—"I propose to apply the expression "artifact" only to the intended products of actions" (1992, p. 60); also—"What is meant by 'a product of an action' in this context? In general, we characterize actions by their results and consequences. The result of an action is some fact or state of affairs brought about by the agent" (Hilpinen 1992, p. 59).

Hilpinen explicitly elaborates a necessary condition that he calls "Dependence Condition" (Hilpinen 1992, p. 65)—"The existence and some of the properties of an artifact depend on an author's intention to make an object of certain kind." Perhaps

[1] Agent is mainly viewed as a unified, self-contained system that interacts with its environment, processes sensory input, and generates behavioral outputs based on internal representations. Importantly, within this framework, the notion of intentionality is central. Intentionality refers to the capacity of the cognitive agent to represent objects, states of affairs, or concepts, and to direct its actions and thoughts towards them. It involves the ability to have mental states that are about something or directed at specific objects or goals. Thus, the cognitive individual agent not only processes information but also exhibits intentional states that guide its behavior and cognition.

the most important component of Hilpinen's definition is that an artifact is an intentionally produced object, and the content of this intention is some type-description of the artifact:

> I have defined an artifact as an intentionally produced object and observed that the content of an intention is some type-description of an artifact. Thus artifacts are essentially and unavoidably instances or tokens of certain culturally determined object-types; this can be regarded as an essential characteristic of artifacts and of many other cultural objects. [...] We can say that an artifact as an artifact is a sign of its culturally determined functional type. (Hilpinen 1992, p. 67)

Artifacts are intentionally produced tokens "of certain culturally determined object-types" (Hilpinen 1992, p. 67). They are tokens of cultural based type-descriptions. For Hilpinen (1992, p. 67), "We can say that an artifact as an artifact is a sign of its culturally determined functional type". We should provide a more detailed explanation of the meaning of this notion, taking into account Peirce's concepts of type-token-tone. The differentiation between a *type* (or legisign) and its *tokens* (or sinsign) is an ontological distinction between a general category of object and its specific, tangible (and replicable) occurrences. According to Savan (1987–1988, p. 13), "Peirce repeatedly indicated that 'laws' have no reality without instantiations in existents. They are regularities in the action of nature that confer on laws their 'being.' That is, laws do not exist independently of their instantiations in a Platonic heaven." In other words, instantiation is a type > token relation (legisign > special type of sinsign). It is the way in which the legisign, a sign that is a law, "governs its replicas," or sinsigns (CP 2.258)—"every legisign signifies through an instance of its application, which may be called a replica of it. The replica is a sinsign. Thus, every legisign requires sinsigns" (CP 2.246). The sinsign, in turn, "involves a qualisign" (CP 2.245), that is, it involves a quality (CP 2.254) or a complex of qualities. "Involvement" is a relation of "implication" of a class of sign, with the character of firstness, in signs of secondness (CP 2.255–262).

For Hilpinen, an artifact is a token of a description (type), which is the content of an agent's intention, simple or complex. There are some problems in this account of the relation between agents and artifacts, when viewed from the perspective of active semiotic externalism. To begin with, the idea of the author as the originator of an artifact puts focus on tracking the causal webs from agents to artifacts. However, tracking these causal webs is hardly straightforward, given that artifacts increasingly are products of complex webs of agency involving multiple individuals across various stages of the production process:

> [T]oday's artworks are often not created by the artists who are their nominal authors. Art is just as likely to be produced by a team of carpenters working onsite; in an enormous fabrication facility; by a luxury goods firm; using a computer program; live; by bodies performing in front of an audience; or by the contributions of hundreds, even thousands, of people, as with Aaron Koblin's *The Sheep Market* (2009), in which over 10.000 worker participants were paid 2 cents to draw a sheep facing left, using Amazon's Mechanical Turk function. (Adamson and Bryan-Wilson 2016, p. 12)

When people and societies apply the notion of authorship, they don't only deal with the causal webs of origination of an artifact, but apply *boundary conditions* to impose some borders on what counts within these causal webs. For example, the name of the master of a workshop, with all the prestige that is associated to them, serves as a boundary condition for delimiting their role as the causal originator of an artwork, despite the many people that may be involved in actually conceiving and producing an artwork in a workshop. Another problem is that Hilpinen's definition of artifact seems to give explanatory primacy to agents with intentions—artifacts are defined in terms of the agents that create them: the psychological phenomenon (intention) is used to explain the material external process, and not the other way around. This seems to go in a direction opposite to that which would be pursued in an emergentist explanation, in which explanatory primacy is given to complex distributed relations involving artifacts. From our point of view, it is more likely that artifacts in a niche may help to explain agents' intentions, than that agents' intentions may help to explain artifacts in a niche. In the case of artworks, the definition of an artifact as an intentionally produced object doesn't resonate well with some central lines of artistic exploration in modern and postmodern art, which target the traditional notion of artist agency and tutorship over creation, such as in the case of *ready-mades*.

10.2 Authorship Is an External Sign

How to conceive of the relation between authors and artifacts according to a process semiotics? According to Hilpinen (1993, pp. 156–157), an artifact must necessarily have an author, which establishes a correlative relationship between the concepts of artifact and authorship—"An object is an artifact if and only if it has an author" (Hilpinen 2011), or multiple authors, and these objects may be referred to as "collectively produced artifacts." As it should be quite clear, we defend a notion of active semiotic externalism, which rejects the explanatory power of the internal/external duality, as seen in internal intention (authorship) determining an external result (artifact). Instead, we will pursue an account of authorship based on an external and self-organized process.

As Nicholson and Dupré insist, in the context of process biology, "things are abstractions from an ever-changing reality. Reality consists of a hierarchy of intertwined processes" (Nicholson and Dupré 2018, p. xi). For Rescher (1996, p. 71), "kinds are constituted as such not because their instances exhibit a commonality in which they *are* but because they exhibit a commonality in what they *do*". This process ontological perspective has important consequences for the notion of authorship of artworks. If artworks are processes whose mode of being is to act—instead of artifacts whose mode of being is to exist as such—authorship has to be viewed as an emergent dynamic based on self-regulation of a process. As Rescher (1996, p. 85) puts it, "physical processes always possess some element of self-propulsion, if not active self-propagation, at least self-perpetuating inertia". We can translate the same argument for the case of sign action. Semiotic processes don't need to be given action by some psychological agent because their mode of being already presupposes agency. If we operate strictly within a process semiotics, there is no such thing as, on the one hand, a passive artifact, a recipient of meaning which is strictly material and non-mental and, on the other hand, an author's mind, which actively imparts meaning and which is strictly psychological. Instead, the artifact and the psychological mind are parts of the same self-organized process of meaning. In Peirce's monistic formulation: "matter is effete mind" (CP 6.24).

Our hypothesis is—semiosis self-organizes, and artifacts, as well as authorship, emerge as results of this self-organization. Self-organization is a process that leads a complex system towards dynamically stable states of increased redundancy and reduced variability (Kelso 1995). In our description, stability is not absence of change (stasis), but regularity through change. Stability is related to the properties of equilibrium and disturbance of a system. Equilibrium is a state that tends to persist. Stability is the capacity of a system to return to equilibrium after a disturbance. We are referring here to the notion of dynamically stable systems, systems whose equilibrium is not static, but depends on an irreversible ordered energy flow to be maintained (Hooker 2011, p. 23). As mentioned in Chap. 2, Peirce's notion of semiosis is compatible with self-organization, especially when we consider how semiotic habits stabilize communication within a system (a community) of utterers and interpreters, through self-correcting dynamics and circular relations within that community (Queiroz and Loula 2011). In this case, the action of signs is more fundamental, in explanatory (as well as ontological) terms, than the intention of an agent. When looking for the agency that brings forth an artwork, we should look at the agency of signs. Understanding semiosis comes before understanding the psychology of authors' minds, the development of cultures, or the representational functions of media artifacts. It is not an internal psychological agency that causes meaning relations. Humans are not only originators of semiosis, but are themselves part of semiosis. Humans themselves are signs enmeshed in real and general semiotic relations (CP 5.314).

How, then, do we investigate the agency of signs? We have to think in terms of distributed agency. Semiosis is temporally distributed, a time-dependent action of signs which develops in several historical and evolutionary scales. In classical examples of distributed cognition, sign action is distributed within spatial contexts: the deck of a ship (Hutchins 1995a), the cockpit of an airplane (Hutchins 1995b), a research laboratory (Nersessian et al. 2003), a theater building (Tribble 2005). In these examples, the distribution of cognition is decisively conceived in terms of spatial *locus*: cognition is not only located in the head of individuals, but in spatial environments of cognitive artifacts and cultural practices. In our approach, the focus of description of distributed sign action is not spatial, but temporal. It is not only the case that a sign cannot simply be described as spatially contained within one's skull, it is also (and more centrally) the case that a sign cannot be described in terms of a locality: it necessitates the notion of a trajectory. Even when the focus of observation is effective sign action instantiated in a single event, the sign acts as a temporally distributed process in which a regular tendency towards certain future outcomes emerges from a history of sign interactions. As explored in the last chapter, the distributed *locus* of sign action is a *niche of semiotic artifacts*.

Artworks and their authorship are distributed processes in a niche. This view has important consequences for the notion of authorship. In this case, authorship is distributed among different brains (neural systems), bodies, artifacts, institutions, and environments. It is in the context of a niche, and subject to the action of initial and boundary conditions that the agency involved in authorship is distributed among agents. There are two processes that participate in the distribution of authorship. First, the emergence of the authorship as a sign in a sociocultural and technical niche. That is, authorship is situated notion that solves problem in some contexts where such problems exist (e.g., authorship may be a sign that solves a problem related to copyright, or yet it may be a sign that solves problems related to individual prestige and recognition; in each case we are talking about different niches that establish conditions for the action of authorship as a sign). Second, a downward determinative influence (Queiroz and El-Hani 2012), when this emergent sign of authorship reorganizes the relation between the components of the system (brains, bodies, artifacts, etc) towards some form of dynamic stability. For example, agents fashion their way of acting towards artifacts so as to embody the authorship sign and produce the expected interpretation, such as being considered authors. Taken together, these two processes constitute the self-organization of authorship. An author is not an originator of an artifact in just any case, but only in a niche where some sort of authorship sign downwardly determines the relation between agent and artifact. Authorship is not an antecedent condition of a semiotic process, but a consequent condition; it is itself an external sign in action.

If authorship is an external sign, it is a *metasemiotic* process (Queiroz and Atã 2019)—it regulates other semiotic processes. In other words, authorship is a sign in action which stands for the action of a sign, or a meta-sign-in-action. If a work of art is a sign in action, authorship is a meta-sign that regulates the action of the artwork as a sign. Knowing the author of an artwork may be central for proper interpretation (production of interpretants) of that work, for instance by allowing an artwork to be connected to other artworks of the same author in a meta-semiotic web of signs that co-regulate each other.

Our main suggestion, in this chapter, is that authorship is an external sign, located not in individual agents, but in niches of semiotic artifacts. This implies that authorship may occupy three specific functional roles, as S, O, or I in S-O-I (see Table 10.1, below). These are complementary views which point to authorship being an irreducibly triadic relation, a genuine sign. If authorship is S in S-O-I, it is a (meta)semiotic artifact (S) that stands for the artwork-agent relation (O) so as to regulate

Table 10.1 An irreducibly triadic version of the relationship between artwork, authorship and interpretation would produce at least three models: (i) artwork determines the interpretation through authorship. (ii) authorship determines interpretation through artwork. (iii) a habit determines authorship (interpretant) through an artwork

If authorship is in…	Then S is…	Then O is…	Then I is…
S	authorship	artwork-agent relation	interpretation
O	artwork	authorship	interpretation
I	artwork	habit	authorship (process)

interpretation of the artwork (I). This is the standard position we assume in our treatment in this chapter. Authorship may also occupy the position of an object (see second line of Table 10.1). This happens, for example, when an interpreter takes an artwork to be a sign of its author, recognizing the author in the artwork, so to speak, or yet a sign of the creation process. Artworks can be seen as standing for aspects of an author's personality, mind or communicative intention, or yet a sign of a creative conception or intention. This may be associated with Hilpinen's view. These two options—authorship as sign of an artwork-agent relation, or artwork as sign of authorship—are not the only analytic alternatives. Authorship can also be seen as occupying the position of interpretant (see third line of Table 10.1 below). In this case, authorship is an interpretative result communicated by a habit through an artwork. Habit, here, is the process of regular action that brings about the artwork (for that interpretant). Authorship is an "interpretation" of that habitual process that an artwork allows to derive. In this case, we are talking of how the interpreter has some power or agency in determining what "authorship" is and who the author (as an immediate object) of an artwork is. This is case in participatory artistic experiments in which the authorship is distributed in the process of producing interpretants of that work.

If authorship is a semiotic process (whether it occupies the place of sign, object, or interpretant), authorship is irreducible to any term of the S-O-I relation considered in isolation from the others. Thus, an implication of our approach is that authorship cannot be located independently (in the sense of dissociation, as discussed in Chap. 2) in the agent, or the artwork, or the interpretation. Authorship can only be considered in the context of an irreducible relation between S-O-I, and is thus better located in the niche, as part of a niche construction process.

10.3 Six Classes of Authorship as a Legisign

If authorship is an external sign, what type of sign is it? Semiosis exhibits a rich variety of morphological patterns. Peircean mature typologies provide a detailed description of several inter-related aspects involved in semiosis including the intrinsic nature of signs and the effect on the semiotic agents (Freadman 2001). That is, each class of sign corresponds to a stable kind of action that takes place between an object, a sign, and an interpretant, and they are classified according to their dynamic relational properties. A sign-action is grounded in some property (qualisign), event (sinsign), or regular pattern (legisign), by virtue of which it stands for some quality (icon), occurrence (index), or law (symbol) to a third element, a possibility (rheme), spatiotemporal connection (dicent sign or dicisign) or rule based tendency (argument) (W 1:332–333).

We suggest that authorship is a legisign, not a quality (qualisign), nor a particular existent (sinsign), but a sign that in itself is a general and regular pattern. A crucial part of our argument, here, consists in a certain way of understanding the type-token relation. Many authors (Currie 1989; Reicher 2015), including Hilpinen (1992), accept the idea that type-token distinction helps to approach the problem of the nature of artworks and the distinction between works that depend more or less on the token (that is, whether and how the type depends on the physical properties of the token). In our approach, the type-token relation needs to be understood as part of a triadic and temporally distributed relation. Type-token relations are only part of a complex of triadically irreducible relations between signs, objects and interpretants.

As a legisign, the object of authorship can be an icon, an index and/or a symbol. It can be an icon if we talk of authorship as standing for a set of qualitative properties (such as the "style" that hypothetically characterizes an authorial agency). It can be an index if we talk of authorship as standing for an existent event, such as a person, a group of persons, an artistic movement (as a historical occurrence). It can be a symbol if we talk of authorship as standing for a general tendency, or habit of action, such as an aesthetic program, a regulatory principle, a formal and/or technical procedure or set of constraints, etc. Finally, the interpretant of "authorship" can be a hypothesis (rheme), a connection (dicent sign or dicisign), or a law (an argument). Thus, authorship can be interpreted as a mere supposition (hypothetical) that some qualitative attributes relate a work to some author; as a proposition (dicisign) that connects a work to an author (such as in the case of a signature in a painting); or yet as a general law that associates some work to an "author-institution" (that is, the author as an institutionalized entity), such as in the case of a legal institution that binds an inherited right of authorial agency (the legisign) to a hereditary line. Another example of authorship as an argument regards the discussion of plagiarism, such as about the borders between what constitutes plagiarism or a creative translation (or transcreation) (de Campos 2007, p. 166). An illustrative example of authorship as legisign is the field of statistical stylistics, or stylometry, which employs different quantitative methods to analyze texts and attribute authorship (Holmes 1998). Statistical stylistics analyzes variables such as word length, sentence length, word frequency, vocabulary richness, etc. to figure out who produced a text, whether literary or not (Clement and Sharp 2003).

A consequence of conceiving authorship as a legisign is that it allows comparing, co-locating, and connecting different aspects of authorship. As a legisign is a type of sign with several different possible and necessary internal relations, viewing authorship as a legisign allows us to account for different aspects of authorship: iconic, indexical and symbolic connections between author(s) and artwork, and rhematic, dicent and argumentative interpretative effects of those connections. This suggests a possible typology of uses of the notion of authorship and how they are connected. Authorship discussed in an investigation of stylistics is probably more based on how the iconic relation of author and artwork produces rhemes, while authorship discussed in a court of law examining copyright infringement is probably is more based on how a symbolic connection between author and artwork produces dicisigns and arguments. These different semiotic relations influence decisively how the notion of authorship is effectively used, discussed, and practiced (Table 10.2).

Table 10.2 Semiotic typology of authorship according to six types of legisigns

Sign class	Description	Example
333	Authorship is interpreted as a matter of declarative rules that typify the connection between author and artifact	A court of law analyses a case of copyright infringement based on how the law typifies an artwork and its author
332	Authorship is interpreted as a direct relation between artifact and agent, but this relation is regulated and typified by rules and habits	A large corpus of poems produced in Brazil during the seventeenth century is directly attributed to the Baroque Brazilian poet Gregório de Mattos, although there is no clear evidence that he wrote all of them
331	Authorship is interpreted as a hypothetical relation between the artifact and a supposed agent	Homer as the author of the Illiad, for those who believe that Homer is a label that expresses a tradition of poetry, rather than an individual
322	Authorship is interpreted as a direct relation between an artifact and an agent (the author is the one who produces the artifact)	João Gilberto is the author of "Bim Bom", because he was the one who wrote the bossa-nova song
321	Authorship is interpreted as a hypothetical direct connection between the agent and the artifact he/she produces	The editors of the New Oxford Shakespeare believe that Marlowe collaborated in Parts 1, 2 and 3 of Henry VI. The conclusion was reached on the advice of an "Attribution Board" of three specialists in the field of literary authorship
311	Authorship is interpreted as a hypothetical set of characteristics shared by artifact and agent	Stylistics discussion to decide whether an artwork can be ascribed to an author

A semiotic typology of authorship as a legisign also has consequences for arts and art history. Different artworks, artists, art movements and historical periods probably focus or address different internal relations of authorship. It could be the case that assignment of authorship in art history is inconsistent, relying heavily on iconic relations in one historical period and on symbolic relations in another, for example. It could also be the case that a person who is considered an "artist", an "author of artworks", is considered so for different reasons in different situations.

10.4 The Agent Is Within Semiosis (Not the Opposite)

From a semiotic externalist perspective, meaning is not within the agent, such as in the form of a communicative intention, but rather the agent is within semiosis. The agent is only an agent within a niche—his/her agency is a distributed property of the niche, which is instantiated in him/her. This is contrary to the view that agency is primarily an intrinsic property within an individual. It is processes of meaning (semiosis) which determine agency, and not agents that determine meaning. This perspective challenges traditional views of authorship that attribute it to the sole and individualistic artist, and highlights the collaborative and networked nature of artistic production involving various actors and factors. For Hilpinen, the agent of authorship is the author, and the content is a culturally determined type, which is also the author's intention. That is, the author acts as an agent who intentionally instantiates types into tokens. In our description, the cultural type is self-organized (an emergent process of niche construction), and this self-organizing process recruits an author, transforming them into an author. Authorship itself can be viewed as a distributed and external legisign-in-action, which is irreducible to particular events (sinsigns) and properties (qualisigns) of individual subjects. This means that authorship is not solely attributed to the artist who physically cause the artwork, but also to the various actors and factors involved in the production and reception of the artwork, such as curators, critics, audiences, and cultural contexts. The author is not necessarily an individual agent but a distributed system, a network, a community. Process semiotics redefines the role of the author, recognizing them as a participant in a dynamic, emergent, and self-organized process of meaning-making, thus reconceiving the role of authors, artworks, and interpreters.

From a process semiotics perspective, we recognize that the author is not a solitary causal originator of authorship. Instead, the author is a *locus* of the action of the authorship sign, which emerges through the interactions between various participants and factors over time. The core of creativity, here, is semiotic emergence (see Chaps. 3 and 4). This emergence is hierarchical in the sense that it is a distributed whole that involves agents and artifacts as indissociable component parts, but it is also diachronic and transformational in the sense that distribution is temporal, based in complex temporal flows between anticipated futures and generalized pasts.

Accepting process (and triadic) semiotic premises dissipates (standard) ideas such as authorship, intentionality, and artifacts. What we mean is that these notions have no explanatory power if we approach the phenomenon of authorship from process semiotic premises. In other words, strict application of a triadic model should modify the metaphysical status of hypothetical entities such as "author", "artifact", "intention", "artwork", reorganizing the metaphysical picture of the phenomenon in terms of temporally-distributed, emergent and self-organized processes.

A feature of our approach is that we describe the whole phenomenon of artwork production and interpretation under the same process semiotic framework. Authorship doesn't require extra explanatory components than semiosis, such as the notions of individual personhood and author intentionality. There is no separation, here, between a class of concepts that describes what goes on within an individual person (psychological states, intentional goals that govern action…) and what goes on outside of him/her (distributed communication mediated by artifacts). An advantage of doing this is that we do away with Cartesian psychological accounts of creativity, and together with it, with any homuncular account of creative agency—there is no creative genius inhabiting an artist and telling them what to do—creative agency has to be studied as a product of niche interactions. What can distinguish authorship from other semiotic processes is not any psychological characteristic inherent to it, but rather the fact that it is a *metasemiotic* process. Here, the author is part of an experiment that semiosis performs upon its own conditions of operation.

Chapter 11
Towards a Typology of Artistic Experimentation: Metasemiotic Experiment as Niche Construction 'Pump'

What is artistic experimentation? What does it experiment upon? How are artistic experiments indeed experiments? This chapter is about the ontology of the artwork. Here, art is defined as a metasemiotic experiment, a process of signs investigating a process of signs. We define this metalevel as a semiotic phenomenon of self-inspection. Artists, like spectators, are situated within self-organized processes. They are not the tutors of semiotic processes but their participants. To structure this argument, based on C.S.Peirce's model of semiosis (action of the sign), we propose a typology—artistic experiments focus on the sign, the object, or the interpretative effects (interpreters) of semiosis, which are the variables they isolate.[1] We illustrate this typology with historical cases from the Rupture Group, artists Cildo Meireles, Berna Reale, and Lygia Clark.

11.1 Art as an Experiment

Here,[2] we suggest a fundamental typology of artistic experiments to answer these questions (What is artistic experimentation? What does it experiment upon? How are artistic experiments indeed experiments?), based on the idea of art as a

The idea of 'intuition pump' appears before Dennett, in Holyoak and Thagard (1994).

[1] Variables in research protocols are attributes or properties that can be isolated, measured, and manipulated. They come in various forms—qualitative or quantitative, continuous or discrete—and can assume both numeric and non-numeric values. For instance, temperature, pressure, and material composition are all examples of variables. An experiment is a controlled test wherein specific factors or variables are altered, while others remain unchanged. This experimental protocol enables researchers to test hypotheses and gain insight into causal relationships between variables.

[2] For a detailed discussion of this topic, see: Livingston (2021).

P. Atã, J. Queiroz, *Mind as an External Sign: A Semiotic Externalist Philosophy of Cognitive Science*, Synthese Library 531,
https://doi.org/10.1007/978-3-032-08988-5_11

metasemiotic experiment. This can be described as a process of accelerating niche construction.[3] In our approach, artistic experiments are experiments concerning the semiotic process itself, focusing on the conditions of emergence, invention, and evolution of semiosis. Artistic experiments are semiotic processes that investigate semiotic processes. This proposition is a generalization, based on C.S. Peirce's model of semiosis as a process, of a thesis explored in various domains—art and literature are investigations of language or investigations of semiotic systems and processes (verbal, visual, volumetric, graphic, auditory, etc.) (Shlovsky 1965; Jakobson and Pomorska 1988; Eco 2002; Noë 2016).[4] In our argumentation, this meta-level is immanent, not transcendent. There is no privileged external entity, such as an agent or an artist, inspecting, monitoring, criticizing, and controlling the semiotic process. Instead, the process self-inspects, self-monitors, self-controls, and self-criticizes. The artist, as well as the audience (spectators or viewers), is situated within the process, which is self-organized. The artist is not the tutor of semiosis but rather integrates into the process as a participant. This "reformulation" (from the artist as an external tutor to the artist as a participant in the process) also constitutes an epistemic and methodological shift with many ontological implications—from the artist as an agent who tutors the process to a network of untutored and distributed interactions among agents, technologies, and different types of cognitive artifacts (Queiroz and Atã 2022). This network can be observed as a self-organized system or process.[5]

The self-investigation of semiotic processes is precisely what constitutes an artistic experiment. Similar to other forms of experimentation, artistic experiments isolate variables within the investigated phenomenon, fixing some while adjusting others to explore how they operate (or hypothetically could operate). The three primary components of the semiotic process are the sign (S) (artwork, material or conceptual artifact, medium, artistic process), the object (O) (the problem, theme, or topic represented), and the interpretant (I) (interpretive effects on the spectators, another sign, action, behavior, or activity).

Based on Peirce's model of categories, there appears to be a hierarchical structure to this model—experiments that concentrate on signs are "foundational" to

[3] Other forms or phenomena of niche construction acceleration include the publication of artistic manifestos, creative translation, critical selection of canon (*paideuma*).

[4] There is a book that should be mentioned - "Art: Strange Tools" by Alva Noë (2016). It draws several parallels between art and philosophical inquiry. Noë argues that both art and philosophy involve a process of exploration, questioning, and investigation. According to Noë, both art and philosophy challenge conventional assumptions, encourage us to question established norms, and explore alternative ways of thinking. Noë also emphasizes that both art and philosophy are dynamic and interactive processes.

[5] Self-organization is a phenomenon primarily observed in complex systems composed of numerous entities that interact mutually, nonlinearly affecting each other's states, leading the system towards an "ordered" state or a state with reduced variability and ambiguity while increasing redundancy (Debrun 2018). Implications of this perspective should encompass a direct connection to a domain of research on complex phenomena grounded in emergent self-organizing processes, their stability conditions, and perturbations.

experiments focusing on sign-object relations, which, in turn, are foundational to experiments examining sign-object-interpretant relations. This hierarchy, as we have seen (Chap. 2), represents a pattern of presupposition and containment of relations—monadic relations (sign) are contained and presupposed by dyadic relations ((sign)object), and these, in turn, are contained and presupposed by triadic relations (((sign)object)interpretant). The sign-object-interpretant relation is thus not a flat relation involving three elements, but a nested organization that can be represented thus (S-O-I(S-O(S))). This typology, in principle, should be able to characterize a significant portion of modern and contemporary art.

In the arts, signs are typically works of art. The object of the work can be a thing, a process, a pattern of activity. It is everything that logically precedes the work and is (or potentially can be) represented by it. Objects are the themes, subjects, events, issues, and problems that a work of art "stands for" and upon which interpretants act. Interpretants encompass all types of effects (conceivable or produced) by the work of art, whether immediate or not. These include the reactions of the public, critical receptions, impact on the art world, influence on artistic movements, translations, adaptations, and recreations. The interpretant is everything that the work of art can determine as a determination of the object through it (or mediated by it).

As we saw (Chap. 2), semiosis is an irreducible process through which a constraining factor (object) influences (determines, regulates) a cognitive behavior (interpretant) mediated by a certain entity, structure, or process (sign). Our approach challenges a well-known assumption: artworks are tools created, employed and/or controlled by artists. One issue with this assumption, as we understand it, lies in its characteristic dualism between the sign and the sign user, and the hierarchy it establishes between them. The level of analysis we employ isn't centered on the sign created, used, or manipulated by the artist, the audience, or the recipient but rather on the semiotic process itself of artistic experimentation. At this level, through the notion of the sign in action (semiosis), we circumvent two explanatory pitfalls—reductionism and reification. In our approach, artistic experiments cannot be reduced to any of their component elements or systems, such as the brain, mental or psychological states, vehicles, media, etc. This property of irreducibility aligns with a general view of semiotic processes as complex and emergent processes. Furthermore, artistic experiments cannot be stripped of a temporal dimension because they unfold in time.

Artworks, artists, audiences, criticism, materials, tools, artifacts, experiments—all are processes and exist only as signs in action. In multi-scale terms, they are processes composing processes. This property of compositionality here follows a principle of fractality or scale invariance—no matter the scale at which we operate, we always observe processes of signs composing processes of signs. Hence, commonly used metaphors to grasp compositionality, such as signs flowing "inside" and "between" artists or "between" artworks and audiences, do not apply. Instead, since artists, artworks, and audiences are participants in semiotic processes, they are all observed as semiotic processes that work together to compose semiotic processes.

Instead of analyzing artistic experiments into component elements, we employ Peirce's method to make analytical distinctions based on component relations, that is, categories or classes of relations. As we have emphasized (Chap. 2), these categories are not classes of substances, entities, or properties. They are classes of relations in action. A semiotic process (or semiosis) is a triadic relation that dynamically connects parts of the world (including the mind, as there is no clear separation between the mind and the world, according to synequism) (EP 2: 646). The process is not defined by these parts but by the triadic relation itself. Thus, a work of art detached from the problems it addresses and the effects it provokes on the spectators cannot be considered a work of art. Another premise is that these elements (artworks, problems addressed and effects on the public) are in a state of flux—a work of art can become a problem or theme addressed by another work of art, just as the effects on a spectator can become the problem or theme of a work of art, or the work itself, and so on. If the object of the work of art is in a triadic relation with the work, it produces an effect that is related to the object through the work. Thus, the work must be determined by the object in relation to the effect and must determine the effect in reference to the object of the work in such a way as to produce the effect to be determined by the object of the work through the mediation of the work.

If there is an irreducibility in the triadic relation (signs, objects, and interpretants), there is also some internal autonomy of internal relations which may be prescinded from the triadic relation. The simultaneous tension between dependence and autonomy makes the semiotic process even more complex. The triadic relationship is hierarchical and asymmetric. It is composed, first, of a monadic relationship of the sign-in-itself; second, a dyadic relationship of the sign with the object; and only third, the triadic relationship with the interpretant. Different artistic experiments isolate (accentuate, concentrate, direct) selective focus on the material or structural properties of the artwork, the relationship between the artwork and the issues or topics it addresses, and/or the production of effects. What we mean by "selective focus" is that when artworks produce effects, some of these effects direct attention to how the artwork produces effects; other experiments direct attention to how the artwork is regulated or determined by the object; others isolate the effect itself. This is a process of self-inspection or self-investigation, in the sense that some part of the process (irreducible relation) examines itself as part of a process (irreducible relation).

The object of the work of art determines the sign, while the interpretant is determined by the determination of the sign by the object. The effect is mediated by the sign, which is the artwork. These three components (sign, object, interpretant) are irreducibly linked, meaning they cannot be conceived outside of the triadic relation itself. They are not defined by any intrinsic ontological attribute, are of a procedural nature, and are self-organized. What artistic experiments do is alter the conditions of the collaborative action (triadic relation) between signs, objects, and interpretants.

11.2 Metasemiotic Experiments

Returning to the initial question, in what sense are artistic experiments truly experiments? Narrower interpretations of the notion of experimentation are associated with very specific conceptions of research and knowledge production practices (Pickering 2016), limiting experiments to the determination of well-defined parameters (Pickering 2016, p. 1) and/or linking experimentation to the evaluation of risky predictive theories (Hacking 1983). In contrast, broader views emphasize how experimental work is not merely an extension or outcome of theorization (Hacking 1983), and/or how experiments radically transform the understanding of research and knowledge production practices (Pickering 2019). Broader views also dissociate experimentation from its more traditionally accepted location, the scientific laboratory.

The promotion of a narrower or broader view of experimentation typically involves a hierarchy of related concepts such as knowledge, belief, science, theory, observation, laboratory, parameter, variable. Depending on the concepts that receive greater emphasis, the notion of experiment takes on a broader or narrower profile of application. In our approach, experiments and experimentation are directly associated with semiotic processes. Hacking (1983, p. 158) suggests that experimental and theoretical work helps to make sense of otherwise "meaningless" observations: "Many phenomena attract great excitement but then they do not yield relevant results because no one can see what they mean, how they connect with anything else, or how they can be used." For Hacking, an experiment is not necessarily preceded or governed by a theory. His criticisms are directed at the Popperian view of experiments as attempts to falsify theoretical claims. Hacking provides several examples of experiments that are not governed by falsifiable theoretical claims and are merely explorations of properties of a certain phenomenon. But if it's not a theory, what motivates and governs exploratory experimental work? There are several examples where the investigation begins with a "notable observation" or simply "noticing some surprising phenomenon" (1983, p. 155). Hacking's framework of scientific investigation as a process initiated through surprise is similar to that advocated by Peirce, for whom scientific investigation is always preceded by surprises. For Peirce, scientific investigation is a process in which an inquirer tries to appease the surprise of an observation by subsuming the surprise under semiotic habits. Surprises and habits are inseparably linked. A surprise implies the breaking of a habit, and the satisfaction of the surprise implies the growth of a habit. A semiotic habit is not a habit of subjects or objects but a habit of signs, or more precisely, a habit of the action of signs. It refers to the stability or regular tendency of certain effects (interpretants) to result from certain causes (objects) due to the mediating action of signs.

What we defend here is that artistic activities or practices consist of experiments on semiotic processes. An artwork is a metasemiotic process. Just as scientific laboratories have the resources (tools and equipment, practices, and suitable environments) to isolate and adjust the variables of an investigated phenomenon, artworks also have the resources to isolate and adjust the conditions for the emergence of semiotic processes. Artworks are metasemiotic laboratories for investigating how semiotic processes emerge and develop under certain conditions. A metasemiotic process (or meta-semiosis) allows for a systematic and critical inspection of semiosis. The artist is an experimental semiotician who produces empirical experiments to generate surprises and surprising effects. These experiments involve variations in conditions, including the isolation and manipulation of variables to produce interpretants. The resulting effects can be logical interpretants, they can affect institutions or specialized critical domains, and public opinion in general. In a literal sense, the public is part of the experiment.

11.3 Fundamental Classes of Artistic Experiment

We have selected five boundary cases, which are operations that isolate the terms of triadic relations. They can be used as prototypical models for our argument that artistic experiments put selective focus either on the sign, the object, or the interpretant of semiosis. The cases presented are the Manifesto of the Ruptura Group (metasemiotic experiment with a focus on S); Cildo Meireles (Coca-Cola Project; experiment with the ready-made object, insertions into ideological circuits, with a focus on O); Berna Reale (Palomo, with focus on O); Lygia Clark and Hélio Oiticica (experiment with active participation of the interpretant, with a focus on I).

11.3.1 Ruptura Group Manifesto: Sign-Dependent Experiment (Dependent on the Qualities and Structures of the Sign)

Grupo Ruptura was a collective of artists that marked the beginning of the concrete art movement in São Paulo, Brazil.[6] Founded in 1952, it was headed by Waldemar Cordeiro (Italian-Brazilian, 1925–1973), who also acted as its main theorist. The founding members were Geraldo de Barros (1923–1998), Luiz Sacilotto (1924–2003), Lothar Charoux (Austrian, 1912–1987), Kazmer Féjer (Hungarian, 1923–1989), Anatol Wladyslaw (Polish, 1913–2004), and Leopold Haar (Polish, 1910–1954). After their first exhibition, Maurício Nogueira Lima, Hermelindo Fiaminghi, and Judith Lauand became part of the group. In their manifesto, they proposed the "renewal of essential values in visual art" through geometric research, bridging the gap between art and industry, and challenging lyrical abstraction as an inappropriate form of individual expression for the artistic context of that era.

[6] See: https://www.tate.org.uk/art/art-terms/g/grupo-ruptura

In the Manifesto (Image 11.1) from 1952, there is a decidedly essentialist formulation about painting. The experiment here focuses on the nature of the sign (S), the material and structure from which S is made. The manifesto makes a distinction between what is "New" and what is "Old," between what it claims as part of its research program and what it rejects and attacks as antithetical to its program. The topics claimed as new are: (1) expressions based on new artistic principles; (2) all experiences that tend toward the renewal of the essential values of visual art (space-time, movement, matter); (3) artistic intuition endowed with clear principles and great practical development possibilities; (4) the attribution of a defined place to art within the framework of "contemporary spiritual work," considering it a means of knowledge that can be deduced from concepts, situating it above opinion, and requiring prior knowledge for its judgment. These topics express an essentialism of the sign (S in S-O-I). Everything considered new here is a perspective on the sign, isolated or dissociated from its object and its interpretant. The second topic expresses this more explicitly by selecting certain aspects—space-time, movement, and matter—as the essential values of visual art. In this semiotic process, the object and the interpretant are subjected to the artifactuality of the sign—space-time, movement, and matter are not means to represent something or to provoke some effect on the audience, but are examined as things in themselves, as independent entities.[7]

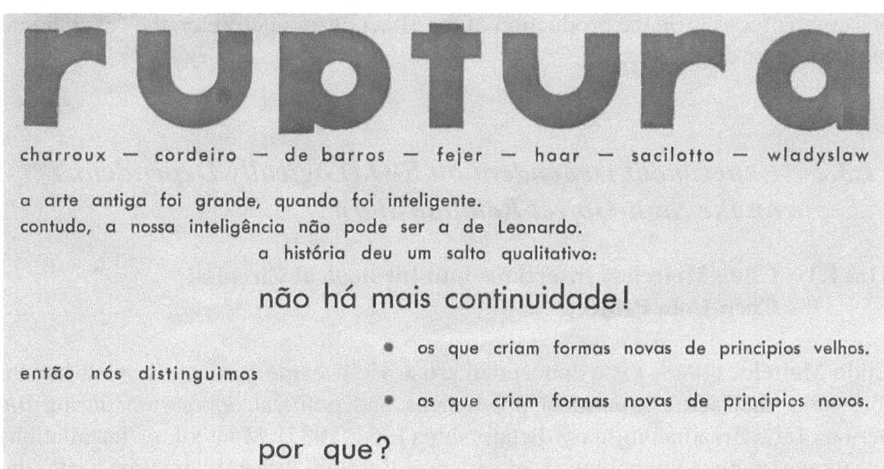

Image 11.1 A fragment of the Ruptura Group Manifesto: "ancient art was great when it was intelligent. / however, our intelligence cannot be like Leonardo's. /history took a qualitative leap. /there is no longer continuity! / Thus we distinguish/ Those who create new forms from old principles / Those who create new forms from new principles/Why?"

[7] It is important to mention that if the experiment is S-dependent, then the qualities that compose S are isolated (manipulated, explored, etc.). However, only specific qualities are targeted—those that give S its ability to function as a sign. S, naturally, possesses numerous physical and material attributes, such as elasticity, density, viscosity, volume, temperature, reflectance, color, smell, etc. The S-dependent experiment isolates certain qualities for experimentation that serve as signs of its object for an interpreter. These qualities form the foundation of the sign.

Another aspect claimed as new is the appeal to clarity above opinion. This claim also expresses an isolation of the sign insofar as it fixes the interpretant. Variations of the interpretant, different opinions, or different ways of synthesizing knowledge are of no interest; what matters is deducibility from constructive principles. The interpreter here is fixed based on a pre-established rationality. There is no room for participatory interpretation, for experimentation with different perspectives, for variations in interpretative framing, for various forms of inference. The interpretant is fixed. Symptomatically, the manifesto considers old (that is, antagonistic to its project) research programs in which objects and interpretants are important: (1) all varieties and hybridizations of naturalism; (2) the mere negation of naturalism, that is, the "wrong" naturalism of children, the insane, primitives, expressionists, surrealists, etc.; (3) hedonistic non-figurativism, a product of gratuitous taste that seeks the mere excitement of pleasure or displeasure. Related to the first two addressed tendencies (naturalism and "wrong" naturalism), they are tendencies in which the action of the object is particularly important. In the case of naturalism, the object is important because it is a mimetic representational trend. In the case of "wrong" naturalism, the object is important because it is a non-mimetic representational trend, that is, a trend in which the sign process fundamentally depends on how the representation of an object can still be representational even if differing from mimesis. As for the third trend, "hedonistic non-figurativism" is attacked precisely because its focus is on the production of the effect on an interpreter, the "excitement of pleasure or displeasure."

11.3.2 Experiment Dependent on S-O (Logically Dependent on the Sign-Object Relationship)

11.3.2.1 Cildo Meireles: Insertions into Ideological Circuits: Coca-Cola Project

Cildo Meireles (1948–) is a conceptual artist. He became particularly well-known for works that address political phenomena and political oppression during the period of the Brazilian military dictatorship (1964–1985). Many of his installations encourage a phenomenological experience through direct interaction with the viewer. Consider, as an example, the Coca-Cola Project (see Image 11.2):

> Meireles conceived his two projects, "Insertions into Ideological Circuits," for a conceptual art exhibition held at the Museum of Modern Art in New York in 1970, titled "Information." The Coca-Cola Project and the Cédula Project explore the idea of circulation and exchange of goods, wealth, and information as manifestations of the dominant ideology. In the Coca-Cola Project, Meireles removed Coca-Cola bottles from normal circulation and modified them by adding critical political statements or instructions to transform the bottle into a Molotov cocktail before returning them to the exchange circuit. Messages such as "Yankees Go Home" appear on the bottles, followed by the artwork's title and the artist's statement of purpose: "Register critical information and opinions on the bottles and return them to circulation." The Coca-Cola bottle is an everyday object of mass circulation. In 1970, in Brazil, it was a symbol of U.S. imperialism and had become glob-

Image 11.2 Insertions into Ideological Circuits: Coca-Cola Project. (Source: Tate Gallery)

ally associated with capitalist consumerism. As the bottle gradually empties of its dark brown liquid, the statement, printed in white letters on a transparent label attached to its side, becomes increasingly invisible, only to reappear when the bottle is refilled for recirculation.[8]

The experiment focuses on the nature of the relationship between the sign and the object and is directly associated with new forms of referentiality. The object of investigation, in this case, is the very S-O relationship of mass consumption objects, such as the Coca-Cola bottle. What is being investigated is not the iconicity of these objects but their referentiality, manipulated to generate surprising effects. The Coca-Cola bottle is typically a sign whose object is the Coca-Cola product or, in the context of that time, a sign of American imperialism. The installation investigates this referentiality, the connection between S-O, and its ability to generate interpretants. The experimental apparatus involves modifying the Coca-Cola bottles themselves, without altering them, to change stable reference relationships. It is crucial for this experiment that the artistic intervention occurs within the index of the themed process. In other words, the political denunciation aimed at American imperialism is supported by bottles that are indexically connected to American imperialism itself. Through this isolation and intervention in the indexical referentiality of the bottle, the referential function of the sign is transformed. The circulation of the consumer product becomes the circulation of subversive ideas. From an index of imperialism, the bottle becomes an index of opposition and resistance to the dictatorship and the imperialism that sustains it.

[8] See: https://www.tate.org.uk/art/artworks/meireles-insertions-into-ideological-circuits-coca-cola-project-t12328

11.3.2.2 Berna Reale: Palomo

Berna Reale, born in 1965 in Belém do Pará, Brazil, is a prominent contemporary artist. She is widely recognized for her performance-based work, which she has been developing since the 1990s. In 2006, she gained significant public attention at the 25th Art Salon of Pará when she presented her work "Cerne." This piece involved a photographic intervention carried out at the Ver-o-Peso meat market, a bustling center for street markets and stalls in Belém, frequented by both tourists and locals.

Wearing a dog's snout and an androgynous police uniform, Reale sits atop a brightly painted red horse and arbitrarily patrols the unusually empty city streets in Palomo (Image 11.3). This image simultaneously mocks the police force and soberly suggests a dormant, lurking, and potentially violent danger (Image 11.4).

All components of S are directly extracted from the referential domains in which S operates. They are represented by S and found in O represented by S. The index of the institution is entirely displayed through its own constituent elements. The object of S is the public safety institution incorporated in the components of S.

The police violence extracted directly from one of its main tools of repression, the police horse, bloodied, ridden by an officer adorned with a muzzle (recommended for difficult-to-control dogs). In one of the most well-known performances, it marches slowly and rhythmically through the deserted city. Police violence, exacerbated by systemic issues, echoes patterns of oppression and human rights abuses that persisted during the dictatorial period after 1964.

Image 11.3 Berna Reale, Palomo (Pigeon), 2012, video still. (Photo of the artist). (https://utah-moca.org/exhibition/berna-reale-singing-in-the-rain/?gad_source=1&gclid=CjwKCAjw7oeqBhBwEiwALyHLM8pl8m8918aIj%2D%2DREVnm3yVq6uVKqYX9K8etO3gkRLNHOya93eQqCBoCGXcQAvD_BwE)

Image 11.4 In the Amazonian city of Belém, Berna Reale's video performance piece "Palomo," created in 2012, is featured in an exhibition in New York. (Image credit: via Berna Reale and Galeria Nara Roesler, São Paulo). (https://www.nytimes.com/2023/04/27/arts/design/brazil-women-artist-exhibition-shiva-gallery.html. https://www.pipaprize.com/2023/05/the-group-show-dissident-practices-explores-how-brazilian-women-artists-respond-to-social-change/)

11.3.3 Experiment Dependent on S-O-I (Logically Dependent on Interpretive Effects)

11.3.3.1 Lygia Clark and the "Bichos"

Lygia Clark (1920–1988) played a pivotal role in the Neoconcrete Group, which she co-founded and where she gained recognition for her experiments with three-dimensional objects displayed in the group's first exhibition in 1959. Clark became especially well-known for works that encouraged active viewer participation. Alongside Hélio Oiticica, she radicalized the Neoconcrete proposal by exploring physical interaction with the artwork. This interactive participation eventually led to the creation of Oiticica's "Bólides" series, which preceded the "Parangolés" and represented a boundary-pushing experience. In the late 1960s and 1970s, Clark developed unconventional projects, such as the participatory series "Bichos" (Image 11.3) from 1960, which consisted of geometric metal constructions articulated through hinges, which were to be manipulated by the participants:

'Bichos.' That's the name I gave to my works from this period because they are fundamentally organic. Furthermore, the hinges that connect the plates make me think of a backbone. The arrangement of the metal plates determines the positions of the 'Bicho,' which at first glance seem limitless. When people ask me how many movements the 'Bicho' can make, I answer: 'I don't know, you don't know, but it knows…'" The 'Bicho' has no reverse side. Each 'Bicho' is an organic entity that fully reveals itself within its inner time of expression. It has an affinity with snails and shells. It is a living organism, a fundamentally active work. A total, existential integration is established between you and it. In the relationship that develops between you and the 'Bicho,' there is no passivity, neither yours nor its. A sort of body-to-body encounter happens between two living entities. In reality, a dialogue takes place in which the 'Bicho' has its own well-defined responses to the viewer's stimuli. This relationship between the human and the 'Bicho,' once metaphorical, becomes real. The 'Bicho' has its own circuit of movements that reacts to the subject's stimuli. It does not consist of independent and static forms that can be manipulated at will and indefinitely, like a game. On the contrary, its parts functionally relate to each other, like those of a true organism, and the movement of these parts is interdependent. In this relationship between you and the 'Bicho,' there are two types of movements. The first, made by you, is purely external. The second, from the 'Bicho,' is produced by the dynamics of its own expressiveness. The first movement (which you make) has nothing to do with the 'Bicho' since it doesn't belong to it. In contrast, the combination of your gesture with the 'Bicho's' immediate response creates a new relationship, and this is only possible thanks to the movements it knows how to make: it is the 'Bicho's' own life.[9]

"Bichos" are mobile, unstable, and entirely manipulable structures. They are constructed both structurally and organically as experiments that rely on the public. The spectator manipulates the sculpture. It is here that the idea of collaboration with the public is presented very clearly in Clark's work for the first time. According to Luis Perez-Orama, "the 'Bichos'… challenge the physical certainty of the user, as they are constantly on the verge of collapse. They have no ideal form. They have no front and back. They are entirely manipulable. And they unfold as potential multiplicities, both structurally and organically."[10]

To investigate the ability of a work of art to acquire dynamic and bio-organic characteristics through its contact with the viewer, the experimental apparatus constructed by Lygia Clark consists of a metal artifact with hinges, capable of resembling the dynamic movements of small animals or insects, but only if these properties are conferred upon the artifact through the integration between the artifact and an active viewer. The sign of the experimental apparatus, therefore, is the articulated "Bicho," and the object is the ability of the "Bicho" to integrate with its interpretant. The interpretant of the experimental apparatus is the viewer's participation, which actualizes (makes real) the relationship between "man" and "Bicho" (Image 11.5).

11.4 Final Comments

This model describes the work of art as a process of investigation into relational properties of semiosis—a metasemiotic process. Its fundamental mode of operation is the isolation of internal semiotic relations: S-dependence, O-dependence, I-dependence. We use the Peircean triadic model as an epistemic tool to describe the work of art as a metasemiotic experiment and propose a typology of its fundamental forms. The work functions as a laboratory, (i) exploring (and inventing) relevant material qualities and dynamic structures, (ii) investigating referential mechanisms

[9] See: https://portal.lygiaclark.org.br/acervo/59268/os-bichos

[10] https://www.moma.org/audio/playlist/181/2429

Image 11.5 Bicho. Source: MOMA (https://www.moma.org/audio/playlist/181/2429)

and processes, (iii) investigating (inventing and expanding) interpretative effects. If the experiment is S-dependent, it isolates (manipulates, explores) the qualities that the work, as a sign, is made of. But not just any qualities. Only those that confer upon S its power to act as S. Of course, S, like any sign, has many physical and material qualities (elasticity, density, weight, viscosity, volume, temperature, reflectance, color, smell, etc.). The S-dependent experiment isolates, for experimentation, those qualities that act in S as a sign of its object to an interpretant. It is the foundation (ground) of the sign that is isolated. The O-dependent experiment isolates the referential connection between sign and object. What matters in this case is to investigate atypical or unconventional possibilities of referentiality, for example when the sign intervenes and becomes part of the object it refers to. If the experiment is I-dependent, the semiotic operation involved in the process of active participation matters, how the interpretant acts on the nature and structure of S, how it modifies S.

We could also state that our model is a kind of method with practical consequences—it provides not just a diagnosis, but also means for identifying what really matters in an artistic experiment, what is most relevant in semiotic terms, and precisely how it should be examined. It functions as a map that allows for precise navigation through various experiments; and it allows for a better specification, in semiotic terms, of what "experimentation" and "research" mean in the arts—the idea of language experimentation.

There is an open problem here, directly influenced by the decision regarding which variables are truly relevant, and how this decision is more or less dependent on the interpreter, the community, and the extent to which there is an objective break from established semiotic habits within the community. Affinities with scientific experimentation are also evident here. In fact, scientific inquiry is the process by which a community of minds attempts to mitigate the surprise of an observation, integrating this surprise into established semiotic habits. Surprises and semiotic habits are two sides of the same coin, oppositional concepts that are inextricably linked. A surprise implies the breaking of a habit, while the appeasement of surprise implies the growth or reinforcement of a habit. In Peirce's semiotics, a semiotic habit is not a habit of subjects or objects, but a habit of signs, or more precisely, a habit of sign action. It refers to the stability or regular tendency of certain interpretative effects (interpretants) to follow from specific causes (objects), due to the mediating action of a sign. If a semiotic habit is a regular action of signs, surprise occurs when something resists this habitual sign action (EP 2:154). Thus, semiotic habits play a role at both the beginning and the end of the process of inquiry—initially, a habit helps define what is surprising, and ultimately, a habit helps to appease the previously formed surprise. In Peirce's pragmatic semiotics, the notion of a semiotic habit is very general and not confined to scientific contexts. For Peirce, "what a thing means is simply what habits it involves" (CP 5.400; Määttänen 2010). Processes of meaning are, in essence, a dynamic negotiation between semiotic habits and surprises. Following Peirce, we understand processes of meaning, the dynamics between habits and surprise, as the foundational background that precedes and motivates experimentation.

Chapter 12
Conclusion

C. S. Peirce has proposed the thesis that mind is semiosis, a triadic process of sign action. 100 years later, after the emergence of process- and complexity-oriented paradigms in cognitive science and evolutionary biology, the thesis that cognition is semiosis takes on new aspects: semiosis is emergent, dynamic, self-organized, multi-level, multimodal (not only symbolic) and out there ("out of our heads").

12.1 Peirce's Triadic Semiosis: A Referential and Interpretive Relationship

Peirce's triadic model of semiosis offers a comprehensive framework and allows us to overcome any binary opposition considering two main tendencies of approaching meaning — sign-object or sign-interpreter-oriented, aligning roughly with referential or interpretative perspectives on meaning. Traditional object-oriented or interpreter-oriented approaches are incomplete because they attempt to reduce the triadic nature of the sign to a dyadic relationship. They either downplay the interpretant, emphasizing the referential relation (sign-object, as seen in semantic externalism), or downplay the object, emphasizing the interpretive relation (sign-interpretant relation, as seen in cognitive internalism).

Peirce's concept of triadic irreducibility asserts that semiosis must be conceived as both sign-object dependent and sign-interpreter dependent (object and interpretant are themselves dynamic parts of a triadic relational process). His intricate model of relations explains how these elements interact to create meaning. In Peirce's framework, a sign mediates between a dynamic object and a final interpretant. These (sign, dynamic object and final interpretant) are the three fundamental phenomenological elements that constitute semiosis, interacting with varying degrees of constraint and/or determination.

P. Atã, J. Queiroz, *Mind as an External Sign: A Semiotic Externalist Philosophy of Cognitive Science*, Synthese Library 531,
https://doi.org/10.1007/978-3-032-08988-5_12

In Peirce's semiotics, the categories of Firstness, Secondness, and Thirdness reflect the degrees of constraint and/or determination by the sign, the dynamic object, and the final interpretant in the semiotic process. Firstness represents the realm of pure possibility and qualities of feeling; it is the immediate and unanalyzed experience. The sign, as the direct experiential element of a semiotic relation, is that which can be perceived, imagined, and/or manipulated. Secondness is the category of actual facts, brute existence, and the reaction between entities. The dynamic object is a determination that compels the sign to produce specific results and not others. Thirdness involves the mediation, laws, habits, and general principles that govern the interactions between the first two categories.

The final interpretant is the coherent generalization of all previous effects produced by the dynamic interpretant through the sign. For example, a diagram of a capybara's anatomy in a biology book is a sign, which is compelled to certain configurations and not others by real capybara that one can point to (dynamical objects), but which ultimately aims at conveying a general class "capybara", with all its associated predicates (the final interpretant). This general class, although it may be exemplified by any actual capybara, is never identical to them. It requires a generalizing inference and an orientation towards future thought, making it impossible to identify in actuality.

Peirce's innovative approach recognizes all these elements as necessary and understands their temporality. Peirce's process semiotics does not argue that meaning is entirely contained in the dynamic object or the final interpretant. The process itself requires the dynamic object and the final interpretant through mediation of the sign, as reflected in the importance of temporal distribution. The meaning of any sign is only contained in its dynamic trajectory. The process of meaning only stands as long as the production of interpretants remains open to the future.

12.2 Emergent Time-Flows: The Creative Temporality of Semiosis

Semiosis is emergent and creative (see Chaps. 2 and 3). In semiosis, relationship between hierarchical levels and diachronic domains of reality is not linear. Following the logic of emergent semiosis, future and past, object and interpretants, don't have to be symmetrical, and there need not be linearity from causes to consequences. In the distributed time of semiosis, the connection between future and past is mediated by generalization of the past and anticipation of the future, with hypothesis (abduction, an ampliative kind of inference) playing an important role in both of these inferential activities. It is such a mediated connection between past and future that allows semiosis to be continuous, but it also makes it probabilistic. To interpret a sign is to discover a temporal trajectory from past to future that regulates that sign. Note that we speak here of interpretation as discovery of tendencies (habits). This is not to say that every interpretant is habitual, but, rather, that if conditions are provided, interpretants tend to develop toward a stable habitual dynamic.

As there are many possible tendencies that may generalize a past and anticipate a future, the distributed time of semiosis is heterochronic, weaved around various emergent temporal flows. A sign may have several different objects and several different interpretants. A semiotic temporal trajectory may develop between a sign and several different objects and interpretants. This proposition embodies, roughly, the ideas of temporal distribution and diachronic emergentism. An event (and the qualities it involves) may also instantiate several different habits. This proposition embodies, roughly, the idea of synchronic emergentism. Thus, for instance, a sinsign may stand for different objects and is an instantiation of different legisigns for different interpretants. This multiplicity is not due to failure in grasping the one truth of a sign, nor due to there being as many truths as there are observers. It is just the nature of a continuous (synechist) process such as semiosis. Interpretative plurality is not an epistemological question, but an ontological one: signs operate in a complex temporality and possess multiple temporal trajectories.

12.3 Semiotic Realism: An Ontology of the Mind

The triadic, non-dualist understanding of mind aligns with Peirce's broader philosophical commitments, such as synechism (the principle of continuity), fallibilism (the idea that no belief is beyond revision), and tychism (the recognition of chance and spontaneity in the universe). These commitments underscore the dynamic and evolving nature of meaning and knowledge. Synechism, in particular, emphasizes that reality is not composed of discrete, isolated entities but is instead a continuum of interrelated processes. Fallibilism asserts that our understanding is always provisional and subject to improvement. Tychism introduces an element of unpredictability, acknowledging that not all events are determined by prior causes.

Peirce was both a scholastic realist and an objective idealist. A Peircean processual approach focuses on the ontology of the semiotic process, in contrast to purely epistemological perspectives, and confers a constitutive role to the interpretant in this ontology. While a realism of substances cannot include the interpretative process (such as observation) as a genuine part of its ontology and relegates it to a secondary, epiphenomenal position, our realism of processes considers the production of interpretants as part of reality. Objects and interpretants cannot be dissociated. Our realism is thus a realism about the semiotic process itself. It is not semiotic realism in the sense that signs refer to real objects but in the sense that semiosis itself is a mode of being.

This is an ontology in which semiotic objects are not substances. In Peirce's philosophy, the notions of fallibilism, synechism, and tychism provide the foundation for understanding the object as a problem. The more a measuring apparatus tries to reach the substantial foundation of a phenomenon, the more it is forced to deal with degrees of indeterminacy: greater precision implies greater indeterminacy, so that the object can only be understood as dynamic, expressed fallibilistically through the comparative generalization between many measurements, and never captured in a single measurement. A main characteristic of a dynamic object is to react in equal force against any attempt to operate an identity between dynamic interpretant and final interpretant. In this sense, the object is, ontologically, a semiotic problem, and cannot be reduced to substantialist "furniture of reality". In this pragmaticist-oriented ontology, reality is relational and needs time. No present point or "slice of time" can be real as such. The mode of being of a dynamic object is to produce a final interpretant in an always advancing future. An object is, ontologically, a process of producing habitual (and thus temporally distributed) semiotic effects in the interpreters that interact with it.

In Peirce's model there is no "final object", but instead a dynamic object and a final interpretant. If objects are problems, the aspiration of inquiry is not to eliminate the problem, reaching an ultimate referent, but rather to develop a stable dynamic between the action of artefacts and reaction of problems, in a process of niche construction. This is our update of Peirce's notion of final interpretant, "which refers to the manner in which the sign tends to represent itself to be related to its Object" (CP 4.536). In our perspective, the final interpretant has a metasemiotic character — at some point, all investigation on the object can only constitute itself further by simultaneously being an investigation on the conditions through which the object produces dynamic interpretant — a metasemiotic investigation. Our examples in art underscore this point. If all inquiry is partly an investigation of the process itself of sign action, in art this character is particularly central — objects, signs and interpretants are inspected in their nature as interwoven components of semiosis.

12.4 A Sign-Centered Externalism

Are mind processes part of the world or part of the agent? They are part of semiosis. Where is the mind? In niches of semiotic artifacts. Are signs mental or material? Peirce's semiotics doesn't start with an examination of non-mental matter or non-material mind. It starts with all that is available to interpretation, anything that may enter in a triadic relation with an interpreter of any nature (social, biological, physico-chemical, computational), a notion that includes everything a dualist would separate into the realms of mind and matter. The traditional divisions of mental/material, internal/external, and cultural/natural are thus challenged. As John Deely (2001, p. 18) argues: "[s]ign is a general mode of being transcending the division of nature and culture".

The externalist position we defend, Active Semiotic Externalism, is indissociable from a process ontological and temporally distributed view — semiosis is cognitively external because it is temporally distributed. Our project here is to semiotize active cognitive externalism. Semiosis is not a process operated in the internal mind of cognitive agents but a process operated in external niches of cognitive artifacts. Thought is in signs and signs are in niches.

Temporal distribution and triadic irreducibility are game changers for the discussion on emergentism. Cognitivist representationalism considers the referent as a substantial entity, and in doing so, it precludes any kind of emergentism not strictly epistemic (weak). If the nature of the referent, or object, is already complete in the object itself, there is little for the interpreter to do other than to refer (truthfully or falsely) to this complete nature. This kind of representationalism eliminates synechist time from semiosis (and pragmatism with it). Here we have the sequential, determinist, time of causation, in which effects are contained in the causes, and are thus ontologically secondary. Interpretations are less real than objective causes, and meaning is not a process but instead a substance. In such a framework, there is no temporal distribution, and thus there cannot be ontological emergence. This is also a dualist approach (for a review of dualist cartesian psychology, see Wheeler 2007) in the sense that the universe is split between a real world of natural substances and a less-than-real world of cognition. Cognition is less-than-real in the sense that either it approximates a real external substance (and thus becomes truthful only by being less semiotic) or it has its membership in the club of reality revoked. In this view, there is no need for real-time in cognition. Dynamical Systems Theory (van Gelder and Port 1991) reintroduces irreversible time in modeling cognition, but it also does away with representation (Brooks 1991; Beer 2000; Chemero 2011), and, in Peircean vocabulary, with Thirdness (generalization, prediction, complex inferential processes…), including temporal distribution and ontological emergence. These approaches are agent-centric, not semiosis-centric.

Discussions on cognitive externalism have been pursued mainly in terms of spatial location: inside or outside the head, spatial extension of the mind, intelligent use of space, and so on. An unusual result of Peirce's thesis is that externalism depends not fundamentally on space, but on time. Temporal distribution is an irreducible part of Peirce's definition of semiosis. Cognition is a temporal relation. Externalism, here, does not only mean displacement in space (out of some head), but displacement in time (out of some succession of temporal slices).

References

Aames, Jimmy. 2021. The concept of the correlate in Peirce's 'New List of Categories'. *Transactions of the Charles S. Peirce Society* 57 (1): 65–88.

Adamson, Glenn, and Julia Bryan-Wilson. 2016. *Art in the making: Artists and their materials from the studio to crowdsourcing*. London: Thames & Hudson.

Atã, Pedro, and João Queiroz. 2014. Iconicity in Peircean situated cognitive semiotics. In *Charles Sanders Peirce in his own words – 100 years of semiotics, communication and cognition*, ed. Torkild Thellefsen and Bent Sorensen, 527–536. Berlin: Walter de Gruyter.

Atã, Pedro, and João Queiroz. 2016. Habit in semiosis: Two different perspectives based on hierarchical multi-level system modelling and niche construction theory. In *Consensus on Peirce's Concept of Habit: Before and Beyond Consciousness*, ed. Donna West and Myrdene Anderson, 109–119. Berlin: Springer.

Atã, Pedro, and João Queiroz. 2019. Semiosis is cognitive niche construction. *Semiotica* 228:3–16.

Atkin, Albert. 2013. Peirce's theory of signs. In *The Stanford encyclopedia of philosophy*, ed. Edward N. Zalta, Summer 2013 Edition. https://plato.stanford.edu/archives/sum2013/entries/peirce-semiotics/. Accessed 08 Apr 2023.

Atkin, Albert. 2016. *Peirce*. London: Routledge.

Atkins, Richard Kenneth. 2018. *Charles S. Peirce's phenomenology*, 2018. Oxford: Oxford University Press.

Baggio, Guido, and Andrea Parravicini, eds. 2019. Symposia. Pragmatism and theories of emergence. *European Journal of Pragmatism and American Philosophy-EJPAP*. https://journals.openedition.org/ejpap/1592

Barker, Clive, and Howard Bay. 2016. Baroque theatres and staging. https://www.britannica.com/art/theater-building/Baroque-theatres-and-staging.

Becvar, Amaya, James Hollan, and Edwin Hutchins. 2008. Representational gestures as cognitive artifacts for developing theories in a scientific laboratory. In *Resources, co-evolution and artifacts*, ed. Mark S. Ackerman, Christine A. Halverson, Thomas Erickson, and Wendy A. Kellogg, 117–143. London: Springer.

Beer, Randall. 2000. Dynamical approaches to cognitive science. *Trends in Cognitive Sciences* 4 (3): 91–99.

Bender, Andrea, and Sieghard Beller. 2017. The Power of 2: How an apparently irregular numeration system facilitates mental arithmetic. *Cognitive Science* 41:158–187.

Bergman, Mats. 2005. C.S. Peirce's dialogical conception of sign processes. *Studies in Philosophy and Education* 24 (3–4): 213–233.

P. Atã, J. Queiroz, *Mind as an External Sign: A Semiotic Externalist Philosophy of Cognitive Science*, Synthese Library 531,
https://doi.org/10.1007/978-3-032-08988-5

Bergman, Mats. 2009. *Peirce's philosophy of communication*. London: The Continuum International Publishing Group.
Bergman, Mats. 2016. Beyond explication: Meaning and habit-change in Peirce's pragmatism. In *Consensus on Peirce's concept of habit: Before and beyond consciousness*, ed. Donna West and Myrdene Anderson, 171–197. Berlin: Springer.
Bergman, Mats, and Sami Paavola. 2020. Tychism. In *The commens dictionary: Peirce's terms in his own words*, ed. Mats Bergman and Sami Paavola. http://www.commens.org/dictionary/term/tychism. Accessed 08 Apr 2020.
Bernstein, Richard J. 2010. *The pragmatic turn*. Cambridge, UK: Polity Press.
Bickerton, Derek. 2009. *Adam's tongue*. New York: Hill and Wang.
Bickhard, Mark. 2003. Process and emergence: Normative function and representation. *Axiomathes*, 14 (1–3): 121–155.
Bjørndahl, Johanne Stege, Riccardo Fusaroli, Svend Østergaard, and Kristian Tylén. 2014. Thinking together with material representations: Joint epistemic actions in creative problem solving. *Cognitive Semiotics* 7 (1): 103–123.
Breton, Gaelle. 1990. *Theater*. Stuttgart/Zürich: Karl Krämer.
Brooks, Rodney A. 1991. Intelligence without representation. *Artificial Intelligence* 47 (1–3): 139–159.
Brunning, Jacqueline. 1997. Genuine triads and teridentity. In *Studies in the logic of Charles Sanders Peirce*, ed. Nathan Houser, Don Roberts, and James Van Evra, 252–270. Indiana University Press: Bloomington.
Buckley, Benjamin Lee. 2012. *The continuity debate: Dedekind, Cantor, Du Bois-Reymond, and Peirce on continuity and infinitesimals*. Boston: Docent Press.
Burch, Robert. 1997. Peirce's reduction thesis. In *Studies in the logic of Charles S. Peirce*, ed. Nathan Houser, Don Roberts, and James Evra, 234–251. Indiana: Indiana University Press.
Cascudo, Luis da Câmara. 2004. *Vaqueiros e Cantadores*. São Paulo: Editora Global.
Clark, Andy. 2004. *Natural-born cyborgs: Minds, technologies, and the future of human intelligence*. Oxford: Oxford University Press.
Clark, Andy. 2005. Word, niche and super-niche: How language makes minds matter more. *Theoria* 54:255–268.
Clark, Andy. 2006a. Language, embodiment, and the cognitive niche. *Trends in Cognitive Sciences* 10 (8): 370–372.
Clark, Andy. 2006b. Memento's revenge: The extended mind, extended. In *Objections and replies to the extended mind*, ed. Richard Menary, 1–43. Oxford: Ashgate.
Clark, Andy. 2008. *Supersizing the mind: Embodiment, action, and cognitive extension*. Oxford: Oxford University Press.
Clark, Andy. 2015. *Surfing uncertainty: Prediction, action, and the embodied mind*. Oxford: Oxford University Press.
Clark, Andy, and David Chalmers. 1998. The extended mind. *Analysis* 58:7–19.
Clement, Ross, and David Sharp. 2003. Ngram and Bayesian classification of documents. *Literary and Linguistic Computing* 18 (4): 423–447.
Currie, Gregory. 1989. *An ontology of art*. London: Macmillan.
Davies, Jim, and Kourken Michaelian. 2016. Identifying and individuating cognitive systems: A task-based distributed cognition alternative to agent-based extended cognition. *Cognitive Processing* 17:307–319.
de Campos, Haroldo. 2007. *Novas: Selected writings of Haroldo de Campos*. Edited by A. S. Bessa and Odile Cisneros. Illinois: Northwestern University Press.
De Tienne, Andre. 2003. Learning qua Semiosis. *S.E.E.D. Journal – Semiotics, Evolution, Energy, and Development* 3:37–53.
De Waal, Cornelis. 2013. *Peirce: A guide for perplexed*. London: Bloomsbury.
Deely, John. 2001. *Four ages of understanding: The first postmodern survey*. Toronto: University of Toronto Press.

Dreyfus, Hubert L. 2002. Intelligence without representation: Merleau-Ponty's critique of mental representation. *Phenomenology and the Cognitive Sciences* 1:367–383.

Dutton, Denis. 1979. Artistic crimes. *The British Journal of Aesthetics* 19 (4): 302–314.

Eaton, Marcia. 1969. Art, artifacts, and intentions. *American Philosophical Quarterly* 6 (2): 165–169.

Engel, Andreas K., Karl J. Friston, and Danica Kragic, eds. 2016. *The pragmatic turn: Toward action-oriented views in cognitive science*. Cambridge, MA: MIT Press.

Esposito, Joseph. 2005. Synechism: The keystone of Peirce's metaphysics. In *The Commens encyclopedia: The digital encyclopedia of Peirce studies*, ed. Mats Bergman and Joao Queiroz, New edition. Retrieved from http://www.commens.org/encyclopedia/article/esposito-joseph-synechism-keystone-peirce%E2%80%99s-metaphysics.

Farias, Priscila, and João Queiroz. 2003. On diagrams for Peirce's 10, 28, and 66 classes of signs. *Semiotica* 147 (1/4): 165–184.

Fisch, Max. 1986. *Peirce, semeiotic and pragmatism: Essays by Max H. Fisch*. Edited by Kenneth Laine Ketner and Christian Kloesel. Indianapolis: Indiana University Press.

Flower, Elizabeth, and Murrey Murphey. 1977. *A history of philosophy in America*. Vol. II. New York: Capricorn Books.

Forster, Paul. 2011. *Peirce and the threat of nominalism*. Cambridge: Cambridge University Press.

Freadman, Anne. 2001. The classifications of signs (II): 1903. In *The Commens encyclopedia: The digital encyclopedia of Peirce studies*, ed. Mats Bergman and João Queiroz. http://www.commens.org/encyclopedia/article/freadman-anne-classifications-signs-ii-1903.

Freeman, Eugene. 1983. *The relevance of Charles Peirce*. La Salle: Monist Library of Philosophy.

Fusaroli, Riccardo, Tommaso Granelli, and Claudio Paolucci, eds. 2011. The external mind. Perspectives on semiosis, distribution and situation in cognition. *Versus: Quaderni di studi semiotici* 112–113.

Gallagher, Shaun. 2009. Philosophical antecedents of situated cognition. In *The Cambridge handbook of situated cognition*, ed. Philip Robbins and Murat Aydede, 35–51. Cambridge: Cambridge University Press.

Gallagher, Shaun. 2023. *Embodied and enactive approaches to cognition*. New edition. Cambridge: Cambridge University Press.

Garland, Ken. 1994. *Mr. Beck's underground map*. London: Capital Transport Publishing.

Gillett, Alex J. 2017, July 6. Week 19: Davies & Michaelian (2016) Identifying and individuating cognitive systems: A task-based distributed cognition alternative to agent-based extended cognition. https://cultureandcognitionreadinggroup.wordpress.com/2017/07/06/week-19-davies-michaelian-2016-identifying-and-individuating-cognitive-systems-a-task-based-distributed-cognition-alternative-to-agent-based-extended-cognition/.

Hausman, Carl. 1993. *Charles Sanders Peirce's evolutionary philosophy*. Cambridge: Cambridge University Press.

Hausman, Carl L. 2002. Charles Peirce's evolutionary realism as a process philosophy. *Transactions of the Charles S. Peirce Society* 38 (1/2): 13–27.

Hilpinen, Risto. 1992. On artifacts and works of art. *Theoria* 58 (1): 58–82.

Hilpinen, Risto. 1993. Authors and artifacts. *Proceedings of the Aristotelian Society* 93:155–178.

Hilpinen, Risto. 1995. Belief systems as artifacts. *The Monist* 78 (2): 136–155.

Hilpinen, Risto. 2011. Artifact. In *The Stanford encyclopedia of philosophy*, ed. Edward N. Zalta. Winter 2011 Edition. https://plato.stanford.edu/archives/win2011/entries/artifact/.

Hoffmann, Michael. 2000. Problems with Peirce's concept of abduction. *Foundations of Science* 4:271–305.

Hoffmeyer, Jesper. 1997. *Signs of meaning in the universe*. Indiana: Indiana University Press.

Holmes, David. 1998. The evolution of stylometry in humanities scholarship. *Literary and Linguistic Computing* 13 (3): 111–117.

Homans, Jennifer. 2010. *Apollo's Angels: A history of Ballet-random House*. New York: Random House.

Hooker, Cliff. 2011. *Philosophy of complex systems*. Amsterdam: Elsevier.

Hookway, Christopher. 1985. *Peirce*. London: Routledge & Kegan Paul.

Hookway, Christopher. 2002. *Truth, rationality, and pragmatism: Themes from Peirce*. Oxford: Clarendon Press.

Houser, Nathan. 1997. Introduction: Peirce as a logician. In *Studies in the logic of Charles Sanders Peirce*, ed. Nathan Houser, Don Roberts, and James Van Evra, 1–22. Bloomington: Indiana University Press.

Hulswit, Menno. 2001. Semeiotic and the cement of the universe: A Peircean process approach to causation. *Transactions of the Charles S. Peirce Society* 37 (3): 339–363.

Hutchins, Edwin. 1995a. *Cognition in the wild*. Cambridge, MA: MIT Press.

Hutchins, Edwin. 1995b. How a cockpit remembers its speeds. *Cognitive Science* 19:265–288.

Hutchins, Edwin. 1999. Cognitive artifacts. In *The MIT encyclopedia of the cognitive sciences*, ed. Robert A. Wilson, 126–128. Cambridge, MA: MIT Press.

Hutchins, Edwin. 2011. Enculturating the supersized mind. *Philosophical Studies* 152:437–446.

Hutchins, Edwin. 2014. The cultural ecosystem of human cognition. *Philosophical Psychology* 27 (1): 34–49.

Iseminger, Gary. 1973. The work of art as artifact. *The British Journal of Aesthetics* 13 (1): 3–15.

Jappy, Tony. 2013. *Introduction to Peircean visual semiotics*. London: Bloomsbury Publishing Plc.

Johnson, Mark, and Steve Larson. 2003. Something in the way she moves: Metaphors of musical motion. *Metaphor and Symbol* 18 (2): 63–84.

Jung, Matthias, and Roman Madzia, eds. 2016. *Pragmatism and embodied cognitive science: From bodily intersubjectivity to symbolic articulation*. Boston: De Gruyter.

Kelso, Scott. 1995. *Dynamic patterns: The self-organization of brain and behavior*. Cambridge, MA: MIT Press.

Kendall, Gary S. 2010. Meaning in electroacoustic music and the everyday mind. *Organised Sound* 15 (01): 63–74.

Kent, B. (1997). The interconnectedness of Peirce's diagrammatic thought. In N. Houser, D. Roberts, & J. Evra (Eds.), *Studies in the logic of Charles S. Peirce* (pp. 445–459). Indiana University Press.

Kim, Jaegon. 1998. *Mind in a physical world: An essay on the mind-body problem and mental causation*. Massachusetts: MIT Press.

Kim, Jaegon. 1999. Making sense of emergence. *Philosophical Studies* 95:3–36.

Kirsh, David. 2009. Problem solving and situated cognition. In *The Cambridge handbook of situated cognition*, ed. Phillip Robbins and Murat Aydede, 264–306. Cambridge: Cambridge University Press.

Kiverstein, Julian, and Michael Wheeler, eds. 2012. *Heidegger and cognitive science*, New directions in philosophy and cognitive science. New York: Palgrave MacMillan.

Kohn, Eduardo. 2013. *How forests think: Toward an anthropology beyond the human*. Berkeley: University of California Press.

Koshkin, Sergiy. 2022. Is Peirce's reduction thesis gerrymandered? *Transactions of the Charles S. Peirce Society* 58 (4): 271–300.

Lakoff, George, and Mark Johnson. 1980/2003. *Metaphors we live by*. London: University of Chicago Press.

Laland, Kevin. 2017. *Darwin's unfinished symphony: How culture made the human mind*. Princeton: Princeton University Press.

Laland, Kevin, J. Odling-Smee, and M. W. Feldman. 2000. Niche construction, biological evolution, and cultural change. *Behavioral and Brain Sciences* 23 (1): 131–175.

Laschi, Cecilia, and Barbara Mazzolai. 2016. Lessons from animals and plants: The symbiosis of morphological computation and soft robotics. *IEEE Robotics & Automation Magazine (September)* 23:107–114.

Lau, Joe, and Max Deutsch. 2019. Externalism about mental content. In *The Stanford encyclopedia of philosophy*. Fall 2019 Edition. https://plato.stanford.edu/archives/fall2019/entries/content-externalism/. Accessed 08 Apr 2023.

Legg, Catherine, and Christopher Hookway. 2019. Pragmatism. In *The Stanford encyclopedia of philosophy*, ed. Edward Zalta. https://plato.stanford.edu/archives/spr2019/entries/pragmatism/. Accessed 08 Apr 2023.

Levinson, Jerrold. 2007. Artworks as artifacts. In *Creations of the mind: Theories of artifacts and their representation*, ed. Eric Margolis and Stephen Laurence, 74–82. Oxford: Oxford University Press.

Liszka, James. 1996. *A general introduction to the semeiotic of Charles S. Peirce*. Bloomington: Indiana University Press.

Lorino, Philippe. 2014. Charles Sanders Peirce, Chapter 10. In *The Oxford handbook of process philosophy and organization studies*, ed. Jenny Helin, Tor Hernes, Daniel Hjorth, and Robin Holt. Oxford: Oxford University Press.

Loula, Angelo, and João Queiroz. 2011. Modeling the emergence and evolutionary history of semiotic systems and processes. *International Journal of Signs and Semiotic Systems* 1:i–iv.

Loula, Angelo, Ricardo Gudwin, Charbel El-Hani, and João Queiroz. 2010. The emergence of self-organized symbol based communication in artificial creatures. *Cognitive Systems Research* 11:131–147.

Määttänen, Pentti. 2010. Habits as vehicles of cognition. In *Ideas in action: Proceedings of the applying Peirce conference*, ed. Mats Bergman, Sami Paavola, Ahti-Veikko Pietarinen, and Henrik Rydenfelt, 201–210. Helsinki: Nordic Pragmatism Network.

Madzia, Roman, and Matthias Jung, eds. 2016. *Pragmatism and embodied cognitive science: From bodily intersubjectivity to symbolic articulation*. Bremen: Walter de Gruyter.

Magnani, Lorenzo. 2005. An abductive theory of scientific reasoning. *Semiotica* 153 (1/4): 261–286.

Magnani, Lorenzo. 2007. Creating chances through cognitive niche construction. In *Knowledge-based intelligent information and engineering systems*, ed. Bruno Apolloni et al., 917–925. Berlin/Heidelberg: Springer.

Magnani, Lorenzo. 2009. *Abductive cognition: The epistemological and eco-cognitive dimensions of hypothetical reasoning*. Berlin/Heidelberg: Springer.

Menary, Richard, ed. 2010. *The extended mind*. Cambridge, MA: MIT Press.

Merrell, Floyd. 1991. Thought-signs, sign-events. *Semiotica* 87 (1/2): 1–58.

Merrell, Floyd. 1994. Of signs and life. *Semiotica* 101 (3/4): 175–240.

Merrell, Floyd. 1996. *Signs grow*. Toronto: University of Toronto.

Merrell, Floyd. 1997. *Peirce, signs, and meaning*. Toronto: University of Toronto Press.

Monteiro, Marianna. 1998. *Noverre: Cartas sobre a Dança*. São Paulo: Edusp.

Murphey, Murray. 1993. *The development of Peirce's philosophy*. Indianapolis: Hackett.

Nersessian, Nancy J. 2008. *Creating scientific concepts*. Cambridge: MIT Press.

Nersessian, Nancy J., Elke Kurz-Milcke, Wendy C. Newstetter, and Jim Davies. 2003. Research laboratories as evolving distributed cognitive systems. In *Proceedings of the Twenty-Fifth Conference of the Cognitive Science Society*, ed. Richard Alterman and David Kirsh, 857–862. Boston: Cognitive Science Society.

Nicholson, Daniel, and John Dupré. 2018. *Everything flows: Towards a processual philosophy of biology*. Oxford: Oxford University Press.

Nielsen, Soeren Nors. 2007. Towards an ecosystem semiotics: Some basic aspects for a new research programme. *Ecological Complexity* 4:93–101.

Noë, Alva. 2015. *Strange tools: Art and human nature*. New York: Hill and Wang.

Odling-Smee, F. John, Kevin Laland, and Marcus Feldman. 2003. *Niche construction: The neglected process in evolution*. Princeton: Princeton University Press.

Oliveira, Luciano. 1999. *A Música na Cantoria em Campina Grande (PB): Estilo Musical dos Principais Gêneros Poéticos*. Thesis, Universidade Federal da Bahia.

Oswalt, Wendell H. 1973. *Habitat and technology: The evolution of hunting*. New York: Holt, Rinehart, and Winston.

Paavola, Sami. 2005. Peircean abduction: Instinct or inference? *Semiotica* 153 (1/4): 131–154.

Paavola, Sami. 2011. Diagrams, iconicity, and abductive discovery. *Semiotica* 186:297–314.

Paavola, Sami. 2012. *On the origin of ideas: An abductivist approach to discovery*. New York: Lap Lambert Academic Publishing.

Pande, Prajakt, and Sanjay Chandrasekharan. 2017. Representational competence: Towards a distributed and embodied cognition account. *Studies in Science Education* 53 (7267): 1–43.

Paolucci, Claudio. 2011. The "external mind": Semiotics, pragmatism, extended mind and distributed cognition. *Versus: Quaderni di studi semiotici* 112–113:69–96.

Parker, Kelly. 1998. *The continuity of Peirce's thought*. Nashville: Vanderbilt University Press.

Peirce, Charles S. 1998. *The essential Peirce: Selected philosophical writings*, Vol. 2 Edited by the Peirce Edition Project. Bloomington/Indianapolis: Indiana University Press. (quoted as EP, followed by volume and paragraph).

Peterson, Donald, ed. 1996. *Forms of representation*. Exeter: Intellect Ltd.

Peterson, A. Townsend. 2011. *Ecological Niches and geographic distributions*. Princeton: Princeton University Press.

Pinker, Steven. 2010. The cognitive niche: Coevolution of intelligence, sociality, and language. *Proceedings of the National Academy of Sciences of the United States of America* 107:8993–8999.

Potter, Vincent. 1997. *Charles S. Peirce: On norms and ideals*. Amherst: University of Massachusetts Press.

Queiroz, João. 2012. Complexification. In *A more developed sign: Interpreting the work of Jesper Hoffmeyer*, ed. Donald Favareau, Paul Cobley, and Kalevi Kull, 67–70. Tartu: Tartu University Press.

Queiroz, João, and Charbel El-Hani. 2006a. Semiosis as an emergent process. *Transactions of the Charles S. Peirce Society: A Quarterly Journal in American Philosophy* 42 (1): 78–116.

Queiroz, João, and Charbel El-Hani. 2006b. Towards a multi-level approach to the emergence of meaning in living systems. *Acta Biotheoretica* 54:179–206.

Queiroz, João, and Charbel El-Hani. 2012. Downward determination in semiotic multi-level systems. *Cybernetics & Human Knowing: A Journal of Second Order Cybernetics, Autopoiesis & Semiotics* 19:123–136.

Queiroz, João, and Sidarta Ribeiro. 2002. The biological substrate of icons, indexes, and symbols in animal communication. In *The Peirce seminar papers – The State of the Art*, ed. Michael Shapiro, 69–78. Oxford: Berghahn Books.

Queiroz, João, and Frederik Stjernfelt, guest editors. 2019. Peirce's extended theory and classification of signs. *Semiotica* 228:1–2.

Queiroz, João, Claus Emmeche, and Charbel El-Hani. 2005. Information and semiosis in living systems: A semiotic approach. *S.E.E.D. Journal: Semiotics, Evolution, Energy, and Development* 5 (1): 60–90.

Queiroz, João, Claus Emmeche, and Charbel El-Hani. 2008. A Peircean approach to information and its relationship with Bateson's and Jablonka ideas. *The American Journal of Semiotics* 24 (1–3): 75–94.

Queiroz, João, Claus Emmeche, Kalevi Kull, and Charbel Nino El-Hani. 2011. The biosemiotic approach in biology: Theoretical bases and applied models. In *Information and living systems – Philosophical and scientific perspectives*, ed. George Terzis and Rob Arp, 91–130. Cambridge, MA: MIT Press.

Quine, Willard Van Orman. 1966. *Selected logical papers*. New York: Random House.

Quine, Willard Van Orman. 1995. Peirce's logic. In *Proceedings of the Charles S. Peirce bicentennial international congress*, ed. Kenneth Ketner et al., 23–31. Texas: Texas Tech Press.

Ramalho, Elba Braga. 2001. Cultura oral em novos estudos: a cantoria nordestina. *O Público e o Privado* 1:7–30.

Ransdell, Joseph. 1977. Some leading ideas of Peirce's semiotic. *Semiotica* 19 (3–4): 157–178.

Ransdell, Joseph. 2003. The relevance of Peircean semiotic to computational intelligence augmentation. *S.E.E.D. Journal: Semiotics, Evolution, Energy, and Development* 3 (3): 5–36.

Reicher, Maria. 2015. *Einführung in die Philosophische Ästhetik*. Darmstadt: wbg Academic.

Rescher, Nicholas. 1996. *Process metaphysics: An introduction to process philosophy*. New York: SUNY Press.

Salthe, Stanley. 1985. *Evolving hierarchical systems: Their structure and representation*. New York: Columbia University Press.

Sartenaer, Olivier. 2016. Sixteen years later: Making sense of emergence (again). *Journal for General Philosophy of Science / Zeitschrift für Allgemeine Wissenschaftstheorie* 47 (1): 79–103.

Sautchuk, João Miguel Manzolillo. 2009. *A Poética do Improviso: Prática e Habilidade no Repente Nordestino*. Thesis, Universidade de Brasília.

Sautchuk, João Miguel Manzolillo. 2011. Poetic improvisation in the Brazilian Northeast. *Vibrant: Virtual Brazilian Anthropology* 8:260–290.

Savan, David. 1986. Response to T.L. Short. *Transactions of the Charles S. Peirce Society: A Quarterly Journal in American Philosophy Summer* XXII (2): 125–143.

Scott-Phillips, Thomas, Kevin Laland, David Shuker, Thomas Dickins, and Stuart West. 2013. The niche construction perspective: A critical appraisal. *Evolution* 68 (5): 1231–1243.

Sharov, Alexei, and Morten Tønnessen. 2021. *Semiotic agency: Science beyond mechanism*. Berlin/Heidelberg: Springer.

Sinha, Chris. 2009. Language as a biocultural niche and social institution. In *New directions in cognitive linguistics*, ed. Vyvyan Evans and Stéphanie Pourcel, 289–310. Amsterdam: John Benjamins.

Sinha, Chris. 2015a. Ontogenesis, semiosis and the epigenetic dynamics of biocultural niche construction. *Cognitive Development* 36:202–209.

Sinha, Chris. 2015b. Language and other artifacts: Socio-cultural dynamics of niche construction. *Frontiers in Psychology* 6:1601.

Spence, Robert. 2007. *Information vizualization: Design for interaction*. Upper Saddle River: Prentice Hall.

Stephan, Achim. 1998. Varieties of emergence in artificial and natural systems. *Zeitschrift für Naturforschung, Tubingen* 53:639–656.

Stephan, Achim. 1999. *Emergenz: Von der Unvorhersagbarkeit zur Selbstorganisation*. Dresden/München: Dresden University Press.

Stjernfelt, Frederik. 2007. *Diagrammatology: An investigation on the borderlines of phenomenology, ontology, and semiotics*. Heidelberg: Springer.

Stjernfelt, Frederik. 2011. On operational and optimal iconicity in Peirce's diagrammatology. *Semiotica* 186:395–419.

Stjernfelt, Frederik, and Markus Pantsar. 2023. Peirce's philosophy of notations and the trade-offs in comparing numeral symbol systems. *Cognitive Semiotics* 16 (2): 121–144.

Sutton, John. 2010. Exograms and interdisciplinarity: History, the extended mind, and the civilizing process. In *The extended mind*, ed. Richard Menary, 189–225. Cambridge, MA: MIT Press.

Thomasson, Amie. 1999. *Fiction and metaphysics*. Cambridge: Cambridge University Press.

Thomasson, Amie. 2004. The ontology of art. In *The Blackwell guide to aesthetics*, ed. Peter Kivy, 78–92. Oxford: Blackwell.

Thomasson, Amie. 2005. The ontology of art and knowledge in aesthetics. *The Journal of Aesthetics and Art Criticism* 63 (3): 221–229.

Tooby, John, and Irven DeVore. 1987. The reconstruction of hominid evolution through strategic modeling. In *The evolution of human behavior: Primate models*, ed. Warren Kinzey, 183–237. Albany: SUNY Press.

Tribble, Evelyn B. 2005. Distributing cognition in the globe. *Shakespeare Quarterly* 56 (2): 135–155.

Vold, Karina, and Schlimm Dirk. 2020. Extended mathematical cognition: External representations with non-derived content. *Synthese* 197 (9): 3757–3777.

West, Donna E., and Myrdene Anderson, eds. 2016. *Consensus on Peirce's concept of habit: Before and beyond consciousness*, Volume 31 of the series Studies in Applied Philosophy, Epistemology and Rational Ethics. Berlin: Springer.

Wheeler, Michael. 2005. *Reconstructing the cognitive world – The next step*. Cambridge: MIT Press.

Whitby, Blay. 1996. Multiple knowledge representations: Maps and aeronautical navigation. In *Forms of representation*, ed. Donald Peterson, 67–79. Wiltshire: Intellect Books.

Zhang, Jiajie. 1997a. Distributed representation as a principle for the analysis of cockpit information displays. *The International Journal of Aviation* 7:105–121.

Zhang, Jiajie. 1997b. The nature of external representations in problem solving. *Cognitive Science* 21:179–217.

Zhang, Jiajie, and Donald Norman. 1994. Representations in distributed cognitive tasks. *Cognitive Science* 18 (1): 87–122.

Zhang, Jiajie, and Vimla Patel. 2006. Distributed cognition, representation, and affordance. *Pragmatics & Cognition* 14 (2): 333–341.

MIX
Papier aus verantwortungsvollen Quellen
Paper from responsible sources
FSC® C105338

If you have any concerns about our products,
you can contact us on
ProductSafety@springernature.com

In case Publisher is established outside the EU,
the EU authorized representative is:
Springer Nature Customer Service Center GmbH
Europaplatz 3, 69115 Heidelberg, Germany

Printed by Libri Plureos GmbH
in Hamburg, Germany